THE NOBEL PRIZE WINNING DISCOVERIES IN INFECTIOUS DISEASES

D1164250

THE NOBEL PRIZE WINNING DISCOVERIES IN INFECTIOUS DISEASES

David Rifkind, MD PhD
and
Geraldine L. Freeman, MD MS

ELSEVIER
ACADEMIC
PRESS

AMSTERDAM • BOSTON • HEIDELBERG • LONDON • NEW YORK
OXFORD • PARIS • SAN DIEGO • SAN FRANCISCO
SINGAPORE • SYDNEY • TOKYO

Elsevier Academic Press
84 Theobald's Road, London WC1X 8RR, UK
http://www.elsevier.com

Elsevier Academic Press
525 B Street, Suite 1900, San Diego, California 92101-4495, USA
http://www.elsevier.com

British Library Cataloguing in Publication Data
A catalogue record for this book is available from the British Library

Library of Congress Catalog Number: 2005925352

ISBN 0 12 369353 5

For information on all Elsevier Butterworth-Heinemann publications
visit our website at www.books.elsevier.com

Typeset by Charon Tec Pvt. Ltd, Chennai, India
www.charontec.com

Printed and bound in Great Britain by Biddles Ltd, Kings Lynn, Norfolk
05 06 07 08 09 9 8 7 6 5 4 3 2 1

Working together to grow
libraries in developing countries

www.elsevier.com | www.bookaid.org | www.sabre.org

ELSEVIER   BOOK AID
            International   Sabre Foundation

CONTENTS

Preface   vii
Acknowledgments   ix

Introduction   1
Alfred Nobel and the Nobel Prizes   5

Part A: Immunity   **11**

  1   Serotherapy   15
  2   Antimicrobial defenses   23
  3   MHC restriction   29

Part B: Antimicrobials   **35**

  4   Prontosil and the sulfonamides   39
  5   Penicillin   43
  6   Streptomycin   47
  7   Chemotherapeutic agents   51

Part C: Bacteria   **55**

  8   Tuberculosis   59
  9   Typhus   67
 10   Syphilis therapy   71

Part D: Viruses   **77**

 11   Tobacco mosaic virus   81
 12   Yellow fever   85
 13   Poliomyelitis virus   89
 14   Hepatitis B virus   95
 15   Bacteriophage   103
 16   Bacteriophage lysogeny   107
 17   Rous sarcoma virus   109

18   Polyoma virus   111
19   Reverse transcriptase   115
20   Viral oncogenes   119
21   Kuru   123
22   Prions   129

Part E: Parasites   **133**

23   Malaria   137
24   Cancer parasite   145
25   DDT   151

Counterpoint: HIV/AIDS   155
Index   161

PREFACE

This book is written for the general reader; it presupposes no special background knowledge in the field of biological sciences. It is intended to be an introduction to the fields of microbiology, immunology and infectious diseases.

A brief biography of Alfred Nobel is presented, with a description of the Nobel Foundation and its prizes. Each of the advances in infectious diseases that have merited recognition by the Nobel Foundation since its inception in 1901 is discussed. They are arranged according to biological groups rather than chronology. Each is presented in the following format: a biographical sketch of the laureate(s), a description of the research, a summary of the current status of the field, and finally some consideration of the relevance of the research to the general field of biology and medicine.

It is not anticipated that every reader will become fully conversant with all of the specific information presented here. Rather the goal of this work is to demonstrate the overall health-related, sociological and economic importance of these prize winning discoveries and to prepare the reader to understand the new infectious disease problems that currently seem to be arising with ever increasing frequency.

ACKNOWLEDGMENTS

This Abstract, which I now publish, must necessarily be imperfect. I cannot here give references and authorities for my several statements; and I must trust to the reader reposing some confidence in my accuracy. No doubt errors will have crept in, though I hope I have always been cautious in trusting to good authorities alone.

(CHARLES DARWIN, 1809–1882)

The primary sources for this work are as follows:

Asimov, I. (1982). *Asimov's Biographical Encyclopedia of Science and Technology*, 2nd edn. Garden City (NY): Doubleday & Company Inc.

Bowker, R. R. (ed.) (1998–1999). *American Men and Women of Science*, 20th edn. New Providence, NJ: Gale Group.

Cook, G. C. and Zumla, A. I. (eds) (2002). *Manson's Tropical Diseases*, 21st edn. Philadelphia, PA: W. B. Saunders Ltd.

Entrez PubMed, available online at: www.nlm\nih.gov. Health Information, Medline/PubMed.

Hoiberg D. H. and Pappas, T. (eds) (2000). *Encylopedia Britannica*, 15th edn. Chicago, IL: Encyclopedia Britannica Inc.

Joklik, E. K., Willett, H. P., Amos, D. B. and Wilfert, C. M. (eds) (1992). *Zinsser Microbiology*, 20th edn. Norwalk, CT: Appleton & Lange.

Knipe, D. M. and Howley, P. M. (eds) (2001). *Field's Virology*, 4th edn. Philadelphia, PA: Lippincott Williams and Wilkins.

Levinovitz, A. W. and Ringertz, N. (eds) (2001). *The Nobel Prize: The First 100 Years*. London: Imperial College Press.

Lodish, H., Berk, A., Zipursky, S. L. *et al*. (2000). *Molecular Cell Biology*, 4th edn. New York, NY: W. H. Freeman and Company.

Mandell, G. L., Bennett, J. E. and Dolin, R. (eds) (2000). *Mandell, Douglas, and Bennett's Principles and Practice of Infectious Diseases*, 5th edn. Philadelphia, PA: Churchill Livingston.

Parker, T. and Topley, W. W. C. (1990). *Topley and Wilson's Principles of Bacteriology, Virology and Immunity*, 8th edn. St Louis, MO: Mosby Year Book.

Strauss, M. B. (ed.) (1968). *Familiar Medical Quotations*. Boston, MA: Little, Brown and Company.

INTRODUCTION

This work presents the twenty-four discoveries in infectious disease that have merited Nobel Prize recognition since the inception of the awards in 1901. Taken as a whole, these discoveries represent a remarkable group of advances and are a tribute to the best of human efforts to create a better life for all of mankind. The Nobel Prizes are instrumental in bringing these benefits to the attention of the world; the extent to which they also may promote and foster further progress can only be a matter of speculation.

It should be recognized, however, that landmark discoveries in infectious diseases were made prior to the establishment of the Nobel awards. Considering the state of scientific knowledge at the time of those early signal advances, the accomplishments were absolutely astonishing. Two prime examples were the introduction of smallpox vaccination by Jenner in 1798 and of rabies vaccination by Pasteur in 1865.

Prior to Jenner's studies, the only method for preventing smallpox was by 'variolation' – a practice originating in China in which lesional fluid from a mild case was inoculated on to the nasal mucosa of the non-immune. Unfortunately this procedure resulted in an occasional smallpox death. Jenner had observed that milkmaids who developed cowpox (vaccinia) lesions on the fingers from infected dairy cattle were subsequently immune to smallpox. Following this lead he inoculated non-immune individuals with cowpox vesicle fluid, and demonstrated that they subsequently became resistant to smallpox. Jenner's agent was the first vaccine devised and is the origin of the term *vaccination*, which is applied both specifically to vaccinia virus as well as generically to any biological substance used to induce immunity. Vaccination with vaccinia virus has resulted in the worldwide eradication of smallpox as of 1980; truly a public health triumph.

1

Rabies has always been and continues today to be an uniformly fatal infection. The World Health Organization estimates that there are 100 000 deaths annually from the infection. In response to the problem, Pasteur prepared a rabies vaccine from desiccated brain and spinal cord of laboratory-infected rabbits and used this vaccine successfully in dog-bite victims with the virus. Fortuitously and uniquely, rabies vaccination can be started after an animal attack because the incubation period of the disease is long – ranging from weeks to months.

At the time of the vaccinia and rabies studies, the discovery of viruses as biological entities, their biological nature and methods of culture were still decades in the future. In addition there was no knowledge of immune mechanisms, and further, Pasteur's 'germ theory of disease' was just being established and accepted. Yet the procedures developed by these two pioneers continue to be used as first-line therapies today.

By the mid-1950s the basic antibiotics used to treat infectious diseases – sulfa, penicillin, streptomycin, the tetracyclines – had been introduced, and now cure a whole range of diseases which previously carried high morbidity and mortality rates. These successes led to a generally held view that infectious diseases as a group were largely under control and no longer posed a major health problem. This optimism has not been borne out by current events, as today infectious diseases remain the leading cause of deaths worldwide.

Coordinated efforts to combat infectious diseases include the Global Fund, supported by the United Nations, which is providing the financial resources to developing countries to combat the major infectious diseases – HIV/AIDS, malaria and tuberculosis. HIV/AIDS has only recently been recognized, while malaria and tuberculosis are diseases of antiquity.

Since the advent of the antibiotic era a number of newly recognized infectious diseases, in addition to HIV/AIDS, have appeared. Examples of these new infectious diseases are numerous and dramatic.

Legionnaires' disease first appeared in 1976 as an outbreak of pneumonia during an American Legion convention in a Philadelphia hotel. The infection was found to be caused by a newly recognized genus of bacteria, which was given the genus name *Legionella*. These bacteria grow in water such as is found in air conditioners, water-cooling towers and similar systems, and are spread by aerosol. The infection is particularly severe in those with depressed immune systems.

In the 1970s and 1980s there were outbreaks in Africa of infections due to two closely related viruses, Marburg and Ebola. These infections are probably transmitted from monkeys to man, and carry a mortality rate of greater than 80 percent.

In 2002 an outbreak of an atypical pneumonia termed Severe Acute Respiratory Syndrome (SARS) occurred in China and, despite strict quarantine measures, rapidly spread to other countries of the world. The outbreak was caused by a corona virus, a class of agents that infect a wide range of animals as well as man. The 2002 outbreak originated in civets, a small arboreal mammal that resides in areas of Asia, Africa and Southern Europe. In the Far East, civets are a culinary delicacy and are sold at high prices in 'wet markets' – shops and stalls dealing in live animals and poultry.

Recently a new type of influenza termed 'bird flu' has appeared in China. The infection has a uniquely high mortality rate and has been traced to domestic chickens. To control the spread of this disease millions of chickens are killed, as the public health community fears that this strain of influenza virus could cross with existing human strains and lead to a major pandemic.

Most recently the US has been subjected to West Nile Virus infections, which cause severe to fatal infections in the central nervous system. This disease is maintained in nature in birds, and is spread by mosquitoes.

The recent mad cow disease in Great Britain and Europe (Chapter 21) and the worldwide pandemic of HIV/AIDS (Counterpoint: HIV/AIDS) are severe medical problems that also have both social and economic consequences.

In addition to these new and exotic microbes, infections due to familiar bacteria and fungi continue to occur due to the development of resistance of the microbes to currently used antimicrobial agents and the use of immunosuppressive agents to treat transplant recipients and cancer patients.

Adding to these problems of natural infections there is now an additional dimension to the challenge of infectious disease: the current threat of biological agents as offensive weapons of terrorism. In the recent past, postal mail containing anthrax bacillus was distributed in the United States, five deaths resulted, and to date the perpetrator remains unknown. Also, although smallpox has been eradicated worldwide, stocks of this virus are maintained by the US and Russian governments. Because smallpox vaccination is no longer routine, the threat of the introduction of smallpox as a weapon of bioterrorism is real. Accordingly, both anthrax and smallpox vaccinations have been reintroduced in the US military as well as in certain first-line emergency response medical personnel. Anthrax and smallpox are not the only potential biological warfare agents; others could also be used.

Finally, it should be noted that the individual Nobel Prize winning discoveries in infectious disease may appear to be isolated independent disclosures; however, when viewed within subject groups, it is apparent that they build progressively one upon the other in a readily discernible continuum. As Isaac Newton commented: 'If I have been able to see further than others it is because I have been standing on the shoulders of giants.'

ALFRED NOBEL AND THE NOBEL PRIZES

Alfred Bernhard Nobel (21 October 1833–10 December 1896) was born in Stockholm, Sweden. His father, Immanuel, was an inventor who sold one of his many devices, a submarine mine, to the Russian government in 1842, and the family moved to St Petersburg so that he could supervise its production of the mine. Here, Alfred received a tutored education. In 1850 he was sent to France for one year of study in chemistry and then on to the United States for four years with John Ericsson, the Swedish-American inventor.

Ericsson had invented the screw propeller for ships, which replaced the paddle wheel, and also built a screw-propelled vessel, the *USS Princeton*, for the navy. During a demonstration of this vessel for high-ranking government officials one of its guns exploded, killing several persons and endangering the life of President John Tyler, who was part of the inspection party. In 1861 Ericsson built the *Monitor*, which battled the Confederacy's *Merrimac* in the first naval engagement between two iron-clad warships, thus ending the reign of wooden naval vessels.

When Alfred returned to Russia in 1855, his father was then engaged in the manufacture of nitroglycerine for use by Russia in the Crimean War. The explosive had been invented by the Italian chemist, Ascanio Sobrero, in 1847. After the war the need for explosives declined, and by 1859 Immanuel's factory was bankrupt. Alfred returned to Sweden and began manufacturing nitroglycerine for peacetime purposes, such as road-building, construction, etc. However, nitroglycerine was a dangerous compound to handle; any shock or jolt would cause it to explode. In

1864 his factory was destroyed by an accidental explosion, killing his brother Emil and four others, and for this reason the Swedish government would not permit the rebuilding of the facility. However, in response to the governmental restriction Nobel improvised a laboratory on a lake barge and there continued his efforts to produce a more stable and manageable form of nitroglycerine. One cask of nitroglycerine in his supply leaked its contents on to the packing material, diatomaceous earth, and Nobel found that the nitroglycerine–diatomaceous earth mixture was stable and could be set off only by a detonating cap – a device which he developed. Nobel named this material 'dynamite', and he manufactured and sold it as formed sticks. He was granted a patent for dynamite in Britain in 1867 and in the United States in 1868. He also invented (and in 1876 patented) 'blasting gelatin', which was dynamite with 7 per cent cellulose nitrate, a product used primarily for underwater work. In addition he invented 'ballistite', a mixture of 40 per cent nitroglycerine and 60 per cent cellulose nitrate used as a solid fuel for rockets. He claimed that his ballistite patent also covered a new explosive, 'cordite', which is essentially ballistite with 5–6 per cent mineral jelly, but this claim was rejected by the British government.

Nobel was a reclusive and self-proclaimed misanthrope who never married, but he considered himself to be a 'super-idealist'. He accumulated great wealth from his explosives patents as well as his holdings in the Baku oil fields in Russia. Upon his death on 10 December 1896 he left his entire estate, valued at about nine million dollars, for the establishment of the Nobel Foundation and its prizes. His will, which he wrote himself, was contested because of both form and content, but after intense litigation and negotiations was approved in 1898.

The will, dated at Paris on November 7, 1895 reads in part:

> The whole of my remaining realizable estate shall be dealt with in the following way: the capital, invested in safe securities by my executors, shall constitute a fund, the interest on which shall be annually distributed in the form of prizes to those who, during the

preceding year, shall have conferred the greatest benefit on mankind. The said interest shall be divided into five equal parts, which shall be apportioned as follows: one part to the person who shall have made the most important discovery or invention within the field of physics; one part to the person who shall have made the most important chemical discovery or improvement; one part to the person who shall have made the most important discovery within the domain of physiology or medicine; one part to the person who shall have produced in the field of literature the most outstanding work in an ideal direction, and one part to the person who shall have done the most or the best work for fraternity between nations, for the abolition or reduction of standing armies and for the holding and promotion of peace congresses ...

In 1900 the Nobel Foundation and a Board of Directors was established, and Nobel Committees were formed to select awardees. This work was supported by the Nobel Institutes.

The Nobel Foundation has the responsibility to maintain the financial base of the awards. Originally the money was invested only in 'safe securities', which at the time referred primarily to bonds. This interpretation has progressively been liberalized so that the full range of bonds, investments, stocks and real estate make up the portfolio.

The original five Nobel Prizes were for Chemistry, Literature, Peace, Physics, and Physiology or Medicine (see table below). In 1905 an additional committee, the Nobel Committee of the Storting (Norwegian Parliament), now known as the Norwegian Nobel Committee, was set up to award the Nobel Peace Prize.

In 1968 the Bank of Sweden established a new award, 'The Central Bank of Sweden Prize in Economic Sciences in Memory of Alfred Nobel', usually referred to as the Nobel Prize in Economics, and in 1969 the first such award was given. After this the Board of the Nobel Foundation decided that no additional Nobel Prizes would be established.

Each award was originally intended for a single individual. At the present, however, an award may be given for two pieces of

Nobel Prizes

Subject	Awarding Institutions
Chemistry	Royal Swedish Academy of Sciences
Economics*	Royal Swedish Academy of Sciences
Literature	Swedish Academy
Peace	Norwegian Nobel Committee
Physics	Royal Swedish Academy of Sciences
Physiology or Medicine	Nobel Assembly at Karolinska Institute

* The Bank of Sweden Prize in Economic Sciences in Memory of Alfred Nobel (established 1968).

research and may be shared among three recipients. While the Peace Prize may be given to institutions, all other prizes are awarded only to individuals.

An individual may not be nominated posthumously, but if the awardee dies after nomination the prize is still awarded. The award consists of a gold medal, a diploma and prize money in an amount depending upon the income of the Foundation. Currently each award amounts to approximately one million dollars. If an award is not given the funds can be held over to the next year, when two prizes can then be given. If not used the second year, the prize money reverts to the fund. If a prize is refused, as occasionally happens for political reasons, the awardee is nevertheless listed as a laureate. That person may later apply, and be given the award, medal, and diploma, but not the monetary prize.

The laureate selection process begins in the autumn of the year preceding the awards. Nominations are received by the committees, followed by evaluations, recommendations to the awarding Institutions, approval, and then the award ceremonies in Stockholm and Oslo in December of the awarding year at a formal dinner and reception hosted by the King of Sweden, with the laureates presenting summaries of their contributions.

Since the first awards were made, in 1901, the Nobel Prizes have been the world's most prestigious recognition of accomplishments – this despite the establishment of other very worthy prizes such as the Lasker Foundation awards for medical science (the 'American Nobel'), the Pulitzer Prize for Journalism, the Booker Prize for Literature, the Fields medal for Mathematics and the Templeton Prize for Studies in Religion. Still, the Nobel remains *non pareil*!

Part A

IMMUNITY

The defense against infectious diseases is provided by the immune system. Any substance that recruits the immune system is termed an *antigen*. Antigens are usually protein or protein–carbohydrate complexes, and the ones most relevant to health are derived from microorganisms, microbial toxins or vaccines prepared from them. In some individual foods, plant pollens or medicines can be antigenic, in which case the result is termed an *allergy*.

There are two components of the immune system, resulting in innate immunity and acquired immunity.

The innate system is present at birth and is the first line of defense against invaders. The innate system is pre-programmed to recognize components that are common among many pathogenic microorganisms. It is comprised of the phagocytic cells, serum complement and miscellaneous other minor systems. The phagocytic cells are found both in the tissues of the body and in the circulation, and protect by ingesting and killing infectious agents. These cells have no antigenic specificity and are always present and available to phagocytose infectious agents.

The complement system is a series of serum proteins which are not antigen-specific and function to increase the efficiency of killing of microorganisms by phagocytic cells and by antibody.

In contrast to the innate system, which is omnipresent, the acquired immune system develops and matures progressively in response to environmental challenges. There are three fundamental characteristics of acquired immunity; lag time, specificity and recall. First, after initial exposure to an antigen there is a delay or lag time of one to two weeks while the immune system gears up to respond. Second, the immune response produced in response to an antigen recognizes and responds only to that specific antigen. Third, if at some future time that same antigen is again encountered, the immune system recalls the initial contact and responds almost immediately without lag-time delay.

The acquired immune system is provided by specialized lymphocytes and is comprised of two limbs, the B-cell limb and the T-cell limb (Figure A.1). The lymphocytes of both the B-cell and the T-cell limbs originate in the bone marrow. The further maturation of the B-cells occurs in the lymph nodes and spleen, and of the T-cells in the thymus.

The B-cell limb is responsible for antibody production and resulting antibody-mediated immunity. Antibody is a class of proteins that is present in the blood serum, in the body tissues and on the mucous membranes. An antibody that is produced in response to an antigen is highly specific for that antigen and no

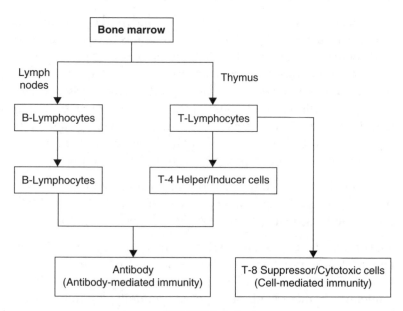

FIGURE A-1
Cells of the acquired immune system.

others. The unique specificity of antibody results from the fact that the molecule has receptor sites that fit exactly to its inducing antigen. The comparison usually made is that the antigen–antibody reaction is like a lock and its key. The specificity of antibody for its inducing antigen rivals the uniqueness of an individual's fingerprints, DNA match and retinal vessel pattern.

The T-cell limb is comprised of two cell types. The T8 suppressor/cytotoxic lymphocytes kill invading microorganisms and infected host cells, resulting in cell-mediated immunity. Similar to the antibody–antigen specificity, the cytotoxic T-cell has receptors that exactly fit its specific inducing antigens on the surface of infected and foreign cells. The T4 helper/inducer lymphocytes assist B-cells in recognizing antigens, and produce antibody.

In antibody-mediated immunity, antibody attaches to microorganisms that multiply in the body outside of host cells and facilitates the killing of those invaders. In cell-mediated immunity, the immune T8 suppressor/cytotoxic cells attach to and kill those host cells in which microorganisms that require an intracellular environment are multiplying.

The critical importance of the acquired immune system in protecting the host from the ubiquitous pathogens in the environment is epitomized by the Severe Combined Immune Deficiency syndrome (SCID). This syndrome is characterized by the total absence of both cell-mediated and antibody-mediated acquired immunity and results in death from any and all infections, usually within the first few months of life. An example was David, dubbed 'the bubble boy' by the popular press. After a previous sibling died of the syndrome, David was tested upon birth, found to be similarly affected and immediately placed in a sterile enclosing capsule where he remained until the age of twelve years. At that time it was felt only humane to release him, following which he developed an infection and died. Currently the SCID syndrome is being treated with bone marrow transplants to provide an intact and functioning immune system.

1

SEROTHERAPY

The diseases of the young are in large part preventable diseases. Epidemics carry off in great proportion the healthy members of a community.

(SIR WILLIAM WITHEY GULL, 1816–1890)

EMIL VON BEHRING

Emil von Behring received the first Nobel Prize for Physiology or Medicine in 1901 for his discovery of antitoxic immunity and the application of this therapeutic modality to diphtheria and tetanus.

Emil Adolf von Behring (1854–1917) was born in Hansdorf, Prussia, obtained his medical degree in 1880, and entered the Army as a surgeon. In 1889 he became an assistant in Robert Koch's Institute of Hygiene in Berlin. He then taught at Halle, Germany, starting in 1894, and in 1895 transferred to Marburg.

DIPHTHERIA

Epidemics of diphtheria, which occur primarily in children, have been reported since ancient times. The term 'diphtheria' derives from the Greek meaning 'leather', and refers to the pseudomembrane that forms in the throat of infected individuals. Death usually results from suffocation and from cardiac toxicity.

The basic studies on the causative bacterium *Corynebacterium diphtheriae* and on the mechanism by which the organism produces disease were carried out in the late nineteenth century. In 1894 Frederick Loeffler, working in Robert Koch's Berlin laboratory, first isolated the bacterium. Initially a puzzling observation was that while the organism could be cultured only from the throat, the infection caused significant heart, nervous system and kidney disease. This observation suggested that the disease process was due to a soluble toxin secreted by the organism which then spread via the bloodstream to distant sites. This conjecture was verified by the demonstration that the disease in guinea pigs could be produced using cell-free filtrates of the organism grown in culture.

Emil von Behring, working in Koch's Berlin laboratory, was the primary investigator in the studies on the prevention and treatment of diphtheria. In 1890 he showed that antisera against the diphtheria toxin could prevent death in experimentally infected animals. The use of antitoxin serum to protect against infection is termed 'passive immunization', as the immune antibody is received from another animal rather than being produced by the immunized animal itself. He then demonstrated that passively immunized animals could be given increasing numbers of toxin-producing organisms, thereby stimulating antitoxin production and leading to 'active immunity'.

In 1913 von Behring showed that children could be actively immunized with infections of the bacterium mixed with antitoxin. This procedure was then widely applied in the prevention of diphtheria outbreaks and epidemics.

It was later shown that an effective vaccine could be produced by treating the toxin with formalin to produce *toxoid*, which was not toxic but still stimulated active immunity. Diphtheria toxoid continues to be used in routine immunizations today.

Occasional cases of diphtheria occur even today in children and adults who have not received the benefit of routine immunization. In these patients the use of antitoxin to produce passive immunity is the basis of effective therapy.

The 1925 diphtheria run

The most famous outbreak of diphtheria in recent times occurred in the winter of 1924–1925 in Nome, Alaska, initially involving about two dozen cases. Curtis Welch, the only doctor in the area, contacted the US Public Health Service in Washington DC for a supply of diphtheria antitoxin. However, the problem of timely delivery of the antitoxin during the winter to Nome seemed almost insurmountable. Air transport in those years was both dangerous and unreliable, and transport by ship to Nome was not possible because the Bering Sea is iced up from October through July. The only option was to use the standard winter service, which was by ship to Seward (an ice-free port) and then north 420 miles via the Alaska Railroad to Nenana and Fairbanks. From Nenana, dog sled teams then traversed west to Nome – a trip of 674 miles that usually required 24 days (Figure 1.1).

It was decided to organize two groups of mushers using volunteers from the small local communities; one group to transport the serum from the railhead at Nenana to Nulato, and the second from Nulato to Nome. Alaska's premier musher, Leonhard Seppala, was in Nome with his experienced lead dog Togo. The serum arrived at Nenana on 27 January 1925, and the first leg, about fifty miles, was taken by Bill Shannon, who passed the supply on to the next team. In the meantime Seppala drove his team from Nome to Shaktoolik, a distance of 170 miles. This required his crossing the ice of Norton Bay in very severe weather, and because several scheduled teams could not participate it was necessary for him to collect the serum and immediately return to Golovin, again crossing the Norton Bay ice. He covered a total distance of 261 miles in contrast to the other teams that went from about 25 to 50 miles each. The final leg from Bluff to Nome, a distance of 53 miles, was driven by Gunnar Kassen, who arrived 1 February. The total distance traveled by dog sled in the serum run of 674 miles was accomplished through the worst weather Alaska could muster, in just five-and-a-half days – in contrast with the usual dog sled time of twenty-four days. The serum effectively aborted the diphtheria epidemic.

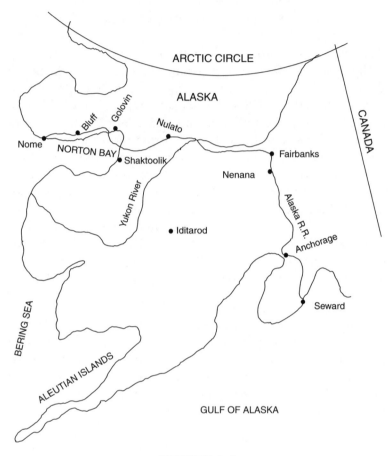

FIGURE 1-1
Map of Alaska showing the route of the 1925 diphtheria run.

There were occasions during the serum run when the wind chill factor dropped to $-70°F$. The critical importance of the lead dog during these times is well recognized by mushers. A not uncommon story is that when the weather deteriorates to the point of zero visibility in blowing snow, the musher has no alternative except to turn the team over to the lead dog and follow along to be taken home safely. Leonhard Seppala and Gunnar Kassen each indicated after the run that he probably could not have survived without the trail skills of his lead dog. The accomplished lead dog is a remarkable animal, almost a breed in and of itself.

On one occasion while crossing the ice of Norton Bay, Seppala's lead dog, Togo, suddenly stopped and literally somersaulted back over the team, causing a snarl of dogs and a tangle of harnesses. Seppala saw that they were only a few yards from a chasm in the ice which would have engulfed team, sled and musher with predictably fatal results.

Gunnar Kassen's lead dog, Balto, is immortalized in a larger than life-size bronze statue near the entrance to the Children's Zoo in Central Park, New York City.

The diphtheria serum run is commemorated annually in March by a sled dog race from Anchorage to Nome billed as 'The Iditarod; The Last Great Race'. The race was originally started in 1967 as a 56-mile run, but as of 1973 the full 1150-mile course has been used. The course does not follow exactly the diphtheria serum route; rather it uses the dog sled mail route established from Knick to Nome in 1910 which goes through the town of Iditarod, taking usually about twenty days.

TETANUS

Emil von Behring also applied his method of serotherapy to tetanus, and detailed those studies in his 1904 book, *The Etiology and Etiological Therapy of Tetanus* and in *The Practical Goals of Blood Serum Therapy*.

Tetanus has figured in medical literature for as long as such literature has existed. It is a distinctive and frightful clinical entity. The disease is characterized by generalized painful muscle spasms involving the entire body, but especially the neck and jaw, hence the term *lockjaw*. The untreated mortality rate exceeds 50 percent. The disease is caused by *Clostridium tetani*. This genus is unique in that the organisms are *anaerobic* – that is, they grow only in the absence of oxygen. In addition, members of this genus produce spores that are highly resistant to heat and drying. These spores occur in the soil and in the intestinal tract of animals and man. Tetanus usually arises following injury suffered in a barnyard, classically from a rusty nail. It is this type of deep puncture wound that the anaerobic organisms require for growth.

When these spores are introduced into the body they revert to the bacterial form and produce a potent toxin, tetanospasmin. This toxin, like diphtheria toxin, is among the most poisonous substances known. One milligram of crystalline tetanus toxin is sufficient to kill approximately fifteen million mice.

The disease is of particular importance in underdeveloped areas, where immunization is not universally practiced. In these areas a form of the disease known as neonatal tetanus occurs, which is due to non-sterile cutting of the umbilical cord. To compound the problem further, in some societies there is a custom of applying manure to the umbilical stump.

Emil von Behring showed that, as with diphtheria toxin, treatment of tetanus toxin with formalin produces a toxoid which immunizes against the disease. He also demonstrated that serum from toxoid immunized horses was useful in treating the disease in man.

Toxoid immunization against tetanus continues to be routine in all developed countries of the world. Booster injections of toxoid are recommended every ten years, but this is often not practiced so that occasional cases occur in adults whose immunity from childhood toxoid immunization has waned. In developing countries, where immunization is not routine, tetanus regularly occurs.

A somewhat unexpected feature of tetanus, and also of diphtheria, is that an infection does not protect against a second attack. This is due to the fact that, because of the extreme potency of the toxins, such small amounts are produced during an infection that the immune system is not activated. Accordingly, after a bout of tetanus or diphtheria, toxoid immunization is required.

SUMMARY

Emil von Behring's work on the serum treatment of diphtheria and tetanus initiated the field of immunology and provided treatment for many forms of human disease, both infectious and non-infectious. The recognition of his contributions with the first Nobel Prize in Physiology and Medicine represented a worthy introduction of the world's most prestigious award in the biological sciences.

2

ANTIMICROBIAL DEFENSES

Blood is a very special juice.

(JOHANN WOLFGANG VON GOETHE, 1749–1832)

JULES BORDET

The 1919 Nobel Prize for Physiology or Medicine was awarded to Jules Bordet for his discovery of factors in the blood that protected against infectious diseases.

Jules-Jean-Baptiste-Vincent Bordet (13 June 1870–6 April 1961) was a Belgian bacteriologist and immunologist who carried out a number of landmark investigations at the Pasteur Institute between 1894 and 1901. In 1901 he founded and directed the Pasteur Institute of Brabant in Brussels, where he collaborated very productively with Octave Gengou. Between 1907 and 1935 he was Professor of Bacteriology at the University of Brussels.

ANTIMICROBIAL DEFENSES

Blood is composed of cells suspended in an aqueous solution of proteins. The cells are red blood cells or *erythrocytes* that carry oxygen, platelets that assist in clotting, and white blood cells. The latter include the granulocytes that ingest microorganisms (a process known as *phagocytosis*) and lymphocytes that provide immunity. The aqueous solution, *plasma*, contains clotting factors, antibody, complement and many other essential components. When the clotting factors are removed, the plasma is then converted to *serum*.

In 1895 Bordet demonstrated that the blood of animals caused lysis of bacteria in the test tube. This bacterial lysis required the presence of two distinct factors working in concert. The first factor appeared in the blood only after prior exposure of the animal to the microorganism, and resisted heating to 56°F for thirty minutes. The second factor was always present, did not require prior exposure to appear, but was destroyed by heating. The heat-stable factor proved to be antibody, while the heat-labile component was later defined as complement.

Antibody

Antibody circulates in the blood serum, specifically in the gamma globulin fraction, and for this reason the terms *antibody*, *gamma globulin*, *immune serum*, and *immunoglobulin* are each used, somewhat colloquially and imprecisely, as synonyms. There are five classes of antibody, each with its own unique molecular structure, distribution in the body and function (Table 2.1). The function of antibody is to attach to infecting microorganisms and to thereby render the microbes susceptible to phagocytosis by host defense cells. In addition, when antibody attaches to microbes the complex then activates the *complement* system.

Immunoglobulin M (IgM), is the class of antibody that is first produced in response to antigen, and this production lasts for several months. Following IgM production there is a switch to immunoglobulin G (IgG), which can continue for years. Because of this sequence, measurement assays of specific IgM and IgG antibodies can reveal whether an infection or immunization is of recent origin, or more remote.

Table 2.1. The five classes of antibody (immunoglobulin).

Designation	Molecular weight	Distribution	Function
IgD	180 000	Surface of B-lymphocytes	Comprises the specific receptors for antigen
IgM	900 000	Circulatory system	Antibody produced for first few months after immunization
IgG	150 000	Circulatory system and body tissues	Antibody class after IgM production has ceased. Continues for years
IgA	160 000* 400 000**	Circulatory system and mucous membranes	Protects mucous membranes against infection
IgE	190 000	Circulatory system, tissues and mucous membranes	Produces allergic reactions to pollens, drugs, foods, etc.

* Form found in serum; ** antibody with protein attached to allow transfer from circulation to surface of mucous membranes.

Immunoglobulin A (IgA) is the class of antibody that protects the mucous membranes against microbial invasion. IgA is produced in the lymph nodes and then linked to a protein, the *transport piece*, so that it can be carried across to the external surface of the mucous membrane. IgA is found in the saliva, tears and mucous lining of the intestinal tract.

Immunoglobulin E (IgE) is found in the tissues and on the mucous membrane. IgE mediates allergic reactions. The positive function of IgE is defense against parasitic round worms, helminths that invade the body.

Immunoglobulin D (IgD) is found on the surface of lymphocytes, where it acts as the receptor for specific antigen. There are at all times an enormous array of small clones of B-lymphocytes, and each is pre-programmed to react to a single specific antigen When the B-lymphocyte contacts its specific antigen, as determined by its surface IgD, it is stimulated to divide, increase in numbers and produce the specific antibody.

An understanding of how an antigenic stimulus results in antibody production was a matter of controversy; focusing around either a directive or a selective mechanism. The directive conceptualization held that the antigen instructed the B-lymphocytes to produce a specific antibody. However, Sir Macfarlane Burnet, in his 1960 Nobel Prize winning studies, showed that the process, rather, was selective. His 'clonal theory of antibody production' holds that there are in the body at all times relatively few B-lymphocytes that are specific for all antigens that might be encountered. Upon contact with an antigen, the small clones of pre-existing B-lymphocytes are induced to replicate and thereby produce large amounts of specific antibody.

Complement

Complement, a term coined by Paul Ehrlich in 1898, is a series of nine serum proteins that aid in the killing of microorganisms. The complement system is activated when antibody attaches to microbes.

The complement proteins function in a sequential series like a line of dominoes that can all be tipped over simply by displacing the first in the series. The cascade is initiated by contact of the first complement protein with an antibody–antigen complex, such as the complex of a microorganism with its specific antibody. Each subsequent complement protein then becomes chemically modified (tipped over), and these modified forms increase both the susceptibility of antibody-coated microorganisms to phagocytosis, and also the blood vessel permeability at the site of infection so that more phagocytic cells and antibody can accumulate locally (see Figure 2.1).

In Figure 2.1, an antigen (such as bacteria) combines with specific antibody, and this combination complexes with a normal

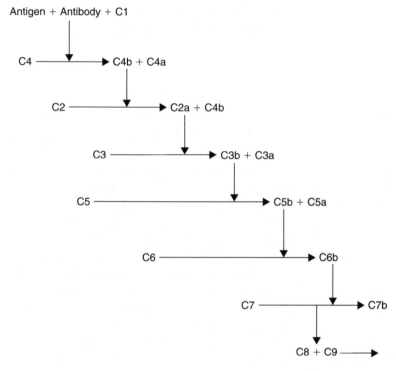

MAC – Membrane Attack Complex

FIGURE 2-1
The complement cascade.

serum protein of the complement system termed C1. This complex then splits another complement system component C4 into C4b and C4a. The C4b is a protein-splitting enzyme and the C4a is a biologically active protein. This cascade continues sequentially ultimately resulting in the formation of the membrane attack complex MAC.

This sequence results in three important antimicrobial defense activities. First, C3b *opsonizes* microorganisms; that is it increases their susceptibility to ingestion by host phagocytic cells. Second, C4a, C3a and C5a increase the permeability of the blood vessels at the site of infection so that increased amounts of antibody and larger numbers of phagocytic cells from the circulation can enter the site of infection. Third, it results in formation of the membrane attack complex which kills microorganisms.

PERTUSSIS

The other major contribution of Bordet was his discovery in 1906 with Gengou of the cause of pertussis. This bacterium, *Bordetella pertussis*, produces in infants and young children a serious disease with respiratory and occasionally central nervous system involvement. The common term, whooping cough, is taken from the paroxysms of coughing followed by a sudden inspiratory whoop which occurs in infected children. This sequence is so characteristic that the experienced physician can diagnose pertussis simply by listening to the child's cough over the telephone. Vaccination of all infants has been routine for more than half a century. The vaccine, prepared from whole killed bacteria, has been controversial because it also can cause fever and vomiting in some cases. An occasional case of permanent brain damage has also been attributed to the whole bacterial vaccine, and it is for this reason that currently vaccination uses an extract of the whole organisms, termed 'acellular' vaccine, which shows less toxicity.

SUMMARY

Emil von Behring's studies on serotherapy disclosed the existence of the immune system and Jules Bordet's analysis of the antimicrobial activity of blood defined the basic components of that system.

3

MHC RESTRICTION

Form ever follows function.

(LOUIS HENRI SULLIVAN, 1856–1924)

PETER DOHERTY AND ROLF ZINKERNAGEL

Peter Doherty and Rolf Zinkernagel shared the 1996 Nobel Prize for Medicine or Physiology for their studies on the mechanism by which the host immune system combats viral infections.

Peter C. Doherty (15 October 1940) is an Australian immunologist and pathologist. He received a bachelor's degree in 1962 and a master's degree in 1966 in veterinary medicine, and a PhD degree in pathology from the University of Edinburgh, Scotland, in 1970. He carried out investigations on lymphocytic choriomeningitis (LCM) between 1972 and 1975, at the John Curtin School of Medical Research in Canberra in collaboration with Zinkernagel. Between 1975 and 1982 he taught at the Wistar Institute in Philadelphia, then from 1982 to 1988 he was Head of the Department of Pathology at the John Curtin School. Since 1988 he has been the Head of the Department of Immunology at St Judes Research Hospital in Memphis, Tennessee.

Rolf M. Zinkernagel (6 January 1944) is a Swiss immunologist and pathologist. He received an MD degree from the University of Basel in 1970 and a PhD from the Australian National University in Canberra in 1975. Starting in 1973, he was a research collaborator with Peter Doherty at the John Curtin School, Canberra. In 1979 he joined the faculty at the University of Zurich, and in 1992 he became Head of the University's Institute of Experimental Immunology.

Doherty and Zinkernagel were interested in whether the nerve damage that accompanies viral infections of the central nervous system results from a direct lethal effect of the virus or possibly from the host immune response intended to eliminate the infection.

THE MAJOR HISTOCOMPATABILITY COMPLEX SYSTEM

There are a number of different viruses that cause *encephalitis* (inflammation of the brain), which is a particularly serious infection because nerve cells (*neurons*) if once damaged do not subsequently repair themselves.

For their research Doherty and Zinkernagel chose, as the laboratory animal model, lymphocytic choriomeningitis (LCM) infection in mice. This is a common worldwide murine infection which occasionally occurs in man, usually after contact with wild mice or pet hamsters.

In their experiments mice were infected with LCM virus to initiate an immune response. Cytotoxic T-cells from LCM-immune mice were mixed in the test tube with mouse cells infected with the virus. The immune T-cells killed the virus-infected brain cells, showing that the host immune response itself could, as well as combating the infection, also contribute to the disease process. An unexpected additional discovery, however, was that this killing occurred only if the immune T-cells and the infected cells both were from genetically identical mice, and more specifically from those having the same major histocompatability complex (MHC). If the immune T-cells and the virus-infected cells were of different MHC types, the infected cells were ignored. This is the essence of *MHC restriction*.

The MHC system was initially recognized as the determinant of the outcome of tissue and organ transplantation. Transplantation was first considered by the Romans to replace gladiators' limbs lost in their contests. In the early twentieth century transplantation was studied in mice using skin grafts. These grafts initially healed, but after a week or two sloughed off (were *rejected*). In a series of studies in the 1950s, Peter Medawar, a 1960 Nobel Laureate, demonstrated that tissue rejection was an immunological process.

In the 1960s and 1970s the genetics of transplantation was defined using skin grafts in various inbred strains of mice, and the controlling system was termed the *major histocompatability complex* (MHC). This system was originally discovered by George David Snell, an American geneticist, who in 1980 received a share of the Nobel Prize in Physiology or Medicine for his studies. The analogous system in man was subsequently defined using human leukocytes, and accordingly termed the *Human Leukocyte Antigen* (HLA) (Figure 3.1). The MHC/HLA

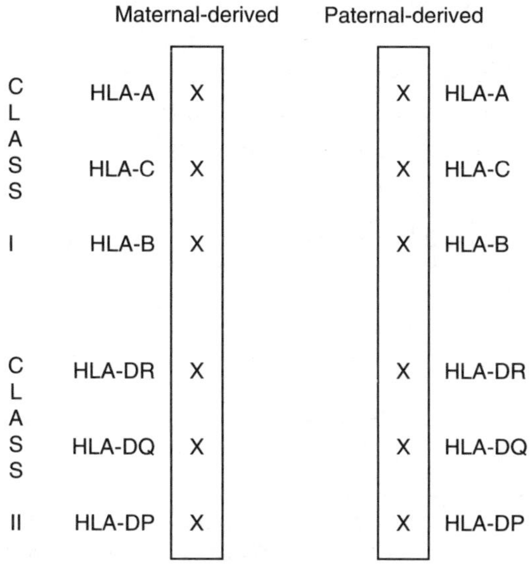

		Maternal-derived			Paternal-derived
C L A S S I	HLA-A	X		X	HLA-A
	HLA-C	X		X	HLA-C
	HLA-B	X		X	HLA-B
C L A S S II	HLA-DR	X		X	HLA-DR
	HLA-DQ	X		X	HLA-DQ
	HLA-DP	X		X	HLA-DP

FIGURE 3-1
The major histocompatability region (chromosome pair #6).

antigens are heterogeneous glycoproteins located on the surface of all body cells. If the MHC/HLA complex of a graft is identical to the recipient's, it is accepted, if different, it is rejected.

In Figure 3.1, the complex on chromosome pair #6 is composed of six Class I and six Class II genes. The three Class I genes on each chromosome are termed HLA-A, HLA-B and HLA-C, while the three Class II genes are HLA-DP, HLA-DQ and HLA-DR. Each of these six gene loci can be occupied by one of a number of alleles (Table 3.1). An allele is a variant form of a specific gene; all variants function similarly but have somewhat different molecular structures. There are hundreds of these HLA alleles; for example at the HLA-A locus there are 220 different alleles available.

Class I HLA antigens are found on the surface of all host cells. Antigens from microorganisms growing within host cells are combined with the HLA antigens. This microbial antigen–HLA complex on the cell surface is contacted by $T8^+$ suppressor/cytotoxic cells which causes the $T8^+$ cells to multiply and disperse, killing microbially infected host cells throughout the body.

Class II HLA antigens are present only on the surface of certain phagocytic cells in the body. Antigens from microorganisms growing extracellularly are taken up by the phagocytic cells; these microbial antigens are complexed with the class II MHC on the cell surface. These MHC–antigen complexes are recognized by

Table 3.1.   Human leukocyte alleles (HLA).

Class	HLA-Gene	Number of alleles*
I	HLA-A	220
	HLA-B	460
	HLA-C	110
II	HLA-DP	116
	HLA-DQ	70
	HLA-DR	360

*Number of alleles that can occupy the genes of the human leukocyte antigen system which is located on the chromosome pair #6. Listed are the number of alleles that can occupy each of the three Class I and three Class II genes. The combinations and permutations of the gene alleles provide the individual's unique HLA signature.

helper/inducer T4-lymphocytes which assist B-cells by means of soluble mediators, termed *interleukins*, in the production of antibody.

The MHC/HLA components are termed antigens because if transferred to another individual they will be recognized as foreign by the recipient's immune system.

The reason for the large number of alleles possibly is that each carries out approximately the same function equally well so that there is no selective pressure to select for one or suppress any other. This is analogous to the fingerprint, where the whorls, ridges and valleys all function equally well to improve the grip on smooth surfaces.

If a person receives a transplanted organ from an individual with a different HLA formula the organ will be rejected. As there are thousands of different possible combinations and permutations of HLA alleles, it is extremely unlikely that a donor–recipient pair will have an identical MHC pattern. It is for this reason that nationwide organ transplant networks are set up to screen and match potential donors and recipients. In general, the closest match is chosen and the recipient given lifelong immunosuppressive therapy to prevent rejection.

SUMMARY

The studies of Doherty and Zinkernagel showed that antigens, such as viruses, do not activate the immune system in their native form. Rather the antigen must first be processed within specialized cells and presented on the surface of these cells in a suitable 'holder'. This 'holder' function is provided by the MHC/HLA molecules. It is only after the immune system encounters the antigen–HLA complex on the surface of the antigen-presenting cell is it activated; and the activation is against the antigen–APC complex, not the antigen alone.

It is obvious that the MHC/HLA system, which has been present in mammals for eons, did not evolve to thwart clinical organ and cell transplantation. It is one of the basic tenets of biology that no system arises and is maintained unless it provides an advantage to the organism. This concept, as expressed by its major exponent, Charles Darwin, is that:

> Natural Selection acts exclusively by the preservation and accumulations of variations, which are beneficial under the organic and inorganic conditions to which each creature is exposed at all periods of life.

It would seem more reasonable that the primary function of the MHC/HLA complex is to present antigen to the immune system. Accordingly, the MHC/HLA complex could more accurately be termed the antigen-presenting complex (APC).

Part B
ANTIMICROBIALS

The antimicrobials include the *chemotherapeutic agents*, which are synthesized in the laboratory, and the *antibiotics*, which are products of microorganisms or plants. Antimicrobials are currently available to treat diseases caused by each of the four major classes of infectious agents: viruses, bacteria, fungi and parasites.

The landmarks in the development and application of antimicrobial agents include the application in the late seventeenth century of cinchona tree bark to treat malaria, the development of sulfas in 1936 and the discovery of penicillin in 1942 to treat bacterial infections, and finally the synthesis of acyclovir in 1970 to treat viral infections.

The mechanisms of action of various antimicrobials include inhibition of the basic metabolic functions common to all living cells; that is, the synthesis of DNA, RNA and proteins. In addition there are antimicrobials that either inhibit the synthesis or alter the permeability of bacterial or fungal cell walls – structures not found in animal cells.

As most of the basic metabolic pathways inhibited by antimicrobials are common to both the microorganisms and the infected hosts, a useful therapeutic effect can be achieved only if the microbial pathways are more susceptible to inhibition than are those in the host cells. The requirement for this differential susceptibility has led to the concept of the *therapeutic index*; a measurement of the relative sensitivities of the microbe and the host to the antimicrobials. A high therapeutic index permits greater treatment efficacy and provides safety in choosing the dosage of an antimicrobial.

There are a few antibiotics, however, that attack a microbial pathway not present in the host – for example, penicillin inhibits the synthesis of the cell wall of bacteria, a structure not present in host cells. Such agents demonstrate a very favorable therapeutic index.

Some antimicrobials kill microorganisms and these are termed *cidal* agents, while others only inhibit the growth of microorganisms and are known as the *static* agents. Intuitively it would seem that the cidal agents would provide the superior therapeutic effect and clinical efficacy; however, except in a few specific diseases, the treatment results are comparable.

Not infrequently two antibiotics in combination are used to treat a bacterial infection. When this therapeutic approach is taken the agents must be chosen with consideration of any unique mutual interaction. There can be *synergism* between the antimicrobial pairs – that is, one plus one is greater than two; an *additive* effect – that is, one plus one equals two; or *antagonism* – that is, one plus one equals less than two. Obviously antagonistic pairs are avoided.

Ever since the introduction of antimicrobials the story has been the launch and application of new effective agents with subsequent appearance of microbial resistance so that efficacy and utility are lost. It seems certain that this alternating process of drug deployment followed by development of antimicrobial resistance will continue into the foreseeable future.

There are four means by which microorganisms acquire resistance to antimicrobials:

1. The target of an antimicrobial can mutate sufficiently so that it is not attacked but still continues to function for the microorganism
2. A bacterium can acquire an enzyme that hydrolyzes and effectively destroys the antimicrobial
3. The outer coat of a bacterium can alter so as to prevent the antibiotic from entering the cell
4. The bacterium can acquire the capacity to expel an antimicrobial rapidly once it has entered the cell so that the agent never achieves an effective response.

The end result of each of these alterations in the bacterium is a development of resistance to the agent.

One method that can be used to delay the appearance of resistance is to use it only where medically necessary. It has been pointed out that there are 130 million prescriptions written annually in the US, and one-half of these are for mild and benign viral infections for which antimicrobials are neither indicated nor effective. Obviously it would be useful to attenuate this practice from the point of view of resistance development as well as its effect in reducing the all-over costs of medical care.

A second source of widespread resistance relates to the large amounts of antibiotics given to domestic animals raised for our food supply. This practice increases the incidence and dissemination of resistant microorganisms and should be discouraged.

The public cannot expect new antimicrobials to be produced indefinitely. On average, it requires about fifteen years to develop, test, and receive FDA approval for a new agent – a process that costs about $900 million. Another factor that limits the development of new antimicrobials is the fact that these agents are usually given on a short-term basis, and therefore the pharmaceutical companies have less incentive to develop an antimicrobial than, for example, a lipid-lowering agent which is taken lifelong.

The development of antimicrobial drug therapy for the treatment of infectious diseases is possibly the most important single advance that has been made in the practice of medicine. However, because of the capacity of microorganisms to become drug resistant, the process of antimicrobial agent development is and must continue to be ongoing. Microbes never sleep.

4

PRONTOSIL AND THE SULFONAMIDES

But in science the credit goes to the man who convinces the world, not to the man to whom the idea first occurs.

(SIR FRANCIS DARWIN, 1848–1925)

GERHARD DOMAGK

Gerhard Domagk was awarded the 1939 Nobel Prize for Physiology or Medicine for his discovery of the antibacterial activity of prontosil. He was prevented from accepting the prize because of Nazi German policy, but in 1947, after the Second World War, he was awarded the gold medal and diploma. The prize money, however, as per the terms of the Nobel awards, reverted to the Foundation.

Gerhard Domagk (30 October 1895–24 April 1964) was a German bacteriologist and pathologist who was in training at the University of Kiel when the First World War began and he entered the German army. He was wounded and was transferred for duty to the Medical Corps. After the war he returned to the University of Kiel and received his medical degree in 1921. He taught at the University of Greifawald in 1924 and the University of Munich in 1925. He then joined the I.G. Farbenindustrie, the German dye company, as director of the Laboratory for Experimental Pathology and Bacteriology, and it was there that he carried out his work on prontosil.

PRONTOSIL

In 1932, workers at the I.G. Farbenindustrie received a patent for prontosil (sulfonamido-crysoldin), a newly synthesized yellow-red azo dye. This new dye attracted Domagk's attention because his main interest was exploring the possible medical therapeutic potential of dyestuffs. His general critique was that as dyes stain bacteria, they might also inhibit their growth and thereby provide a therapeutic effect in infectious diseases. Accordingly, in conjunction with an ongoing examination of over a thousand azo dyes, Domagk tested prontosil against streptococcal infection in mice and found the compound to be effective.

Domagk used prontosil in 1935 to treat a serious streptococcal infection in his young daughter, Hildegarde, who recovered promptly. The drug was also used by Franklin D. Roosevelt Jr, the President's son, again with positive results. Its use in 1935 in a 10-year-old girl with hemophilus meningitis unfortunately was not curative. However, the therapeutic use of the drug in that patient brought prontosil to the attention of the medical and scientific communities.

A somewhat paradoxical feature of prontosil was that, while effective in treating infections in animals, it had no antibacterial effect in the test tube. This phenomenon was analyzed by Daniel Bovet, a Swiss pharmacologist who headed the therapeutic chemistry unit at the Pasteur Institute, Paris. He reasoned that in the body prontosil might be converted to an antimicrobially active form, and accordingly in 1936 he tested some of the breakdown products of the molecule in the test tube and found one of them, sulfanilamide, to be the active factor. Parenthetically, Bovet received the 1957 Nobel Prize for his studies on antihistamines and muscle relaxants.

THE SULFAS

The mechanism of action of the sulfas was worked out by P. Fildes and D. D. Woods in the early 1940s – the 'Fildes–Woods hypothesis'. The synthesis of nucleic acids requires folic acid and its derivatives, dihydrofolic acid and tetrahydrofolic acid (Figure 4.1). In bacteria, the folic acid series is synthesized from pteridine plus para-aminobenzoic acid (PABA). As sulfa structurally resembles PABA, it competitively inhibits its incorporation into folic acid.

In man, folic acid is a vitamin – that is, it cannot be produced in the body but must be acquired in the diet. It is this dietary-acquired folic acid that is used in the synthesis of DNA, and accordingly this synthesis in man is not inhibited by sulfa.

Sulfonamides were the first antimicrobial agents used during the Second World War, preceding the introduction of penicillin (see Chapter 5) by about five years. Since then a number of different sulfonamide drugs have been synthesized, with differing pharmacologic characteristics, and applied in clinical medicine.

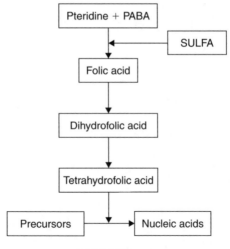

FIGURE 4-1
The mechanism of antimicrobial action of the sulfonamides.

SUMMARY

With the introduction of new antimicrobials during the past fifty years, the use of the sulfonamides has declined. However, they still have specific indications, especially in combination with other inhibitors of microbial metabolism. Furthermore, the search to understand the mechanism of action of antibiotics presaged the important studies of Hitchings and Elion (Chapter 7), who refined and extended this approach to the development of more advanced antimicrobial agents.

5

PENICILLIN

In the fields of observation chance favors only the
prepared mind.

(LOUIS PASTEUR, 1822–1895)

ALEXANDER FLEMING, ERNEST CHAIN AND HOWARD FLOREY

The 1945 Nobel Prize for Physiology or Medicine was shared among Alexander Fleming, Ernest Chain and Howard Florey for their discovery of penicillin and for devising methods for purification and production of the antibiotic for clinical use.

Sir Alexander Fleming (6 August 1881–11 March 1955) was a Scottish bacteriologist who, after graduation from Kilmarnock Academy, worked in London as a clerk. He joined the Army in 1900, and in 1902 earned a scholarship to the St Mary's Hospital Medical School, University of London, from which he graduated in 1906. He served in the British Army's Medical Corps during the First World War. In 1919 he continued his research at St Mary's, where he became Hunterian Professor in 1919, and Arris and Gale Lecturer at the Royal College of Surgeons in 1928. He was elected a fellow of the Royal Society in 1943, and knighted in 1944.

Sir Ernest Boris Chain (19 June 1906–12 August 1979) was a German-English biochemist who obtained his PhD in chemistry

43

and physiology from the Friedrich Wilhelm University in Berlin in 1930 and then did research at the Institute of Pathology, Charité Hospital, Berlin, from 1930 to 1933. He then went to the University of Cambridge and two years later to the University of Oxford, where he worked with Florey on penicillin. From 1948 to 1961 he was Director of the International Research Centre for Chemical Microbiology, Superior Institute of Health, Rome. He returned to London and from 1961 to 1973 he was Professor of Biochemistry, then from 1973 to 1976 he was Professor Emeritus and Senior Research Fellow, and from 1978 to 1979 he was a Fellow. He was knighted in 1969.

Baron Howard Walter Florey (24 September 1896–21 February 1968) was an Australian-English pathologist who obtained his MD degree from the University of Adelaide in 1921. As a Rhodes scholar he traveled to England and studied at Oxford, following which he attended Cambridge and received a PhD degree in 1927. He then taught pathology, first at the University of Sheffield in 1931, and then at Oxford beginning in 1935. He was knighted in 1945, elected President of the Royal Society in 1960 and Provost of Queen's College, Oxford, in 1960. In 1965 he was given a life peerage and became Baron Florey of Adelaide.

PENICILLIN

Fleming's primary interest was in the chemotherapy of infectious diseases. He introduced into Great Britain Ehrlich's arsphenamine (Neosalvarsan, '606'), an arsenic-containing drug used to treat syphilis. His major contribution was the recognition, in 1928, that a mold produced an antibacterial agent. During studies on variants of staphylococcus he had left a culture plate open on his St Mary's Hospital laboratory bench, and noted that there was a zone of bacterial inhibition around an airborne fungus that had fallen on the plate. This fungus, later named *Penicillium notatum*, was characterized and found to be closely related to the common bread mold. He recognized that the product of this mold, which he named penicillin, was not toxic to white blood cells and therefore had potential therapeutic utility, but he did not immediately pursue the discovery. However, with the advent of the Second World War the need for treatment of soldiers with infected wounds became a priority, and to this end Chain and Florey set about production of Fleming's penicillin.

Florey was able to isolate penicillin from broth culture of the fungus, and Chain developed a chemical assay test for the antibiotic. Together they developed deep broth culture techniques for its large-scale production.

In the initial stages the antibiotic preparation was only 10 percent pure. In 1941 nine patients were treated in England with penicillin, with very encouraging results. At the time it required about twenty-five gallons of broth culture to produce sufficient antibiotic to treat one patient for one day. Administered penicillin is excreted unchanged in the urine, and the supply was so scarce that urine from the patient was collected and the excreted penicillin recovered and reused for treatment. Recycled penicillin! These early patients were treated with remarkably small doses of penicillin – from 3000 to 5000 units daily. In succeeding years, doses up to 100 million units were occasionally used.

Fermentation procedures for production were rapidly improved, and in 1943 in North Africa the antibiotic was first used to treat wartime casualties. In this same year penicillin was approved for use by the US military. By 1950, 150 tons of penicillin per year were being produced, and currently a day's dosage of penicillin costs only pennies.

Penicillin is an almost a perfect antibiotic because it inhibits the synthesis of the bacterial cell wall – a component not present in man. For this reason there is not a deleterious effect on the host, so the tolerable dosage is almost unlimited. The only limitation with penicillin is the occurrence of allergies of several types. These, although relatively rare, can be serious, and occasionally fatal.

Resistance

In the 1960s, however, infections unresponsive to penicillin began to appear, due to bacteria which had previously been sensitive to the antibiotic. The bacteria had become resistant because they had acquired the capacity to produce an enzyme, penicillinase, which was originally discovered by Chain. The enzyme caused the breakdown of the penicillin molecule. To counter this developing resistance a new series of penicillins, termed 'semisynthetic', were developed and produced. In these antibiotics the penicillin nucleus was obtained by the usual penicillium culture procedure and then chemically modified to produce derivatives resistant to the enzyme penicillinase. A number of these derivatives were produced and are currently in use.

In addition to this approach a second strategy was developed; that is, the combination of penicillin with a second agent also of microbial origin that inhibited penicillinase. Penicillin plus penicillinase inhibitors represents a class of antibiotics in widespread use today.

6
STREPTOMYCIN

cornucopia {a late L. form, written as one word of an earlier *cornu copial* 'horn of plenty'; fabled to be horn of the goat amalthea by which the infant Zeus was suckled; the symbol of fruitfulness and plenty.}

The horn of plenty; a goat's horn represented in art as overflowing with flowers, fruit, and corn.

An overflowing stock or store.

(THE OXFORD ENGLISH DICTIONARY)

SELMAN WAKSMAN

Selman Waksman was a soil microbiologist who received the 1952 Nobel Prize for Physiology or Medicine for his discovery of the antibiotic, streptomycin.

Selman Abraham Waksman (22 July 1888–16 August 1973) was an Ukrainian-born soil microbiologist and biochemist who emigrated to the US in 1910 after high school and was naturalized in 1916. He graduated from Rutgers University, New Jersey, in 1915, and then studied at the University of California, Berkeley, receiving a PhD degree in 1918. He returned to Rutgers, where he was Professor of Soil Microbiology from 1930 to 1940, Chairman of the Microbiology Department from 1940 to 1958, and

Director of the Rutgers Institute of Microbiology from 1949 to 1958.

Among the many different soil microorganisms, Waksman was particularly interested in the class of 'higher bacteria' which are characterized by growing in long filamentous chains, thus placing them phylogenetically between the typical bacteria and the fungi or molds. Waksman, with his associates, isolated three important products from higher bacteria.

STREPTOMYCIN

In 1940 he isolated and crystallized actinomycin from *Streptomyces parvullus*. While this compound displayed antimicrobial activity, it was too toxic for routine use in treating infections. However, it was found to be active against certain tumors of childhood as well as some uterine and testicular tumors. In addition, actinomycin had some utility in the management of the immunologic rejection process in kidney transplant recipients.

Waksman's major contribution was the discovery of streptomycin, which he isolated from *Streptomyces griseus*. It was with the discovery of this compound that he coined the term 'antibiotic'. While penicillin is active primarily against Gram-positive organisms, streptomycin is active against a wide variety of both Gram-positive and Gram-negative bacteria. Uniquely it was shown to be active against the tubercle bacillus, and it was in fact the first antibiotic to show significant activity in the treatment of tuberculosis.

An important general question in microbiology in the 1950s was answered using streptomycin. It was known that if bacteria were cultured in the presence of an antibiotic, some of the organisms would become resistant to it. The question was whether the antibiotic specifically directed the appearance of specifically resistant organisms, or whether resistant variants arose spontaneously and were simply selected for by the presence of the antibiotic.

The question was explored in the following experiment. One hundred tubes of streptomycin-free broth were inoculated with bacteria and incubated simultaneously. Each of these cultures was then transferred individually to streptomycin-containing agar plates and incubated, and the streptomycin-resistant colonies were counted. The plates showed a wide variation in the number of colonies that appeared, ranging from none to hundreds. Logically, if streptomycin were inducing resistance then all of the plates would have had approximately the same number of resistant colonies. The experimental result, however, proved that resistant cells arose spontaneously and at random times in the absence of the antibiotic. If the resistant cells arose early in the incubation of the original plain broth tubes, they would multiply so that many colonies would appear on the streptomycin-containing plates. In contrast, if the resistant mutants arose late in the course of incubation then only a few resistant colonies would appear on the streptomycin-containing agar plates.

This experiment, showing that resistance arose spontaneously and was not specifically directed, had important implications not only in microbiology but also in genetics in general. All mutations are now held to be random events. Mutagenic agents (such as X-rays) do not direct specific changes in the chromosomal DNA; rather they simply increase the rate at which these spontaneous disruptions occur.

It is of passing historical interest that the major proponent of the directed mechanism of mutation was Sir Cyril Hinshelwood, a British chemist who in 1956 shared the Nobel Prize in Chemistry for his work on the physical chemistry of the water molecule.

The clinical use of streptomycin has declined in recent years. The antibiotic, in high doses, can produce significant ear toxicity, resulting in both loss of hearing and loss of balance. Newer antibiotics in the same chemical family as streptomycin, the aminoglycosides, have now come online and are in routine clinical use.

Subsequently, in 1949, Waksman isolated another aminoglycoside antibiotic, neomycin, from *Streptomyces fradiae*. While too toxic for systemic administration, it is one of the antibiotics in an ubiquitous drugstore triple antibiotic combination ointment used to treat minor infections in scratches and abrasions.

SUMMARY

The lasting importance of Waksman's research is not in the specific antibiotics that he discovered. Rather his approach – the search for medically useful products in soil microorganisms – continues to be used today by the pharmaceutical industry and has contributed many therapeutic agents now in routine clinical use.

7

CHEMOTHERAPEUTIC AGENTS

A discovery is rarely, if ever, a sudden achievement, nor is it the work of one man; a long series of observations, each in turn received in doubt and discussed in hostility, are familiarized by time, and lead at last to the gradual disclosure of truth.

(SIR BERKELEY MOYNIHAN, 1865–1936)

GEORGE HITCHINGS, GERTRUDE ELION AND JAMES BLACK

The 1988 Nobel Prize for Physiology or Medicine was shared by George Hitchings and Gertrude Elion for their improved method for designing new therapeutic agents.

George Herbert Hitchings (18 April 1905–27 February 1998) was an American pharmacologist who received bachelor's and master's degrees from the University of Washington, Seattle, and in 1933 received a PhD degree in biochemistry from Harvard University. He taught at Harvard until 1939, and then from 1942 to 1975 he conducted research at the Wellcome Research Laboratories, Burroughs Wellcome Company, Research Triangle Park, North Carolina.

Sir James Whyte Black (14 June 1924) is a British pharmacologist who was director of therapeutic research at the Wellcome Research Laboratories.

Gertrude Belle Elion (23 January 1918–21 February 1991) was an American pharmacologist who graduated from Hunter College, New York City, in 1937 with a degree in biochemistry. While searching for a research position she taught high school chemistry until 1944, when she obtained employment at the Wellcome Research Laboratories. She was initially a research assistant, then a colleague of Hitchings – a collaboration that lasted four decades.

CHEMOTHERAPEUTIC AGENTS

Prior to the contributions of Hitchings, Elion and Black, the classical procedure for finding therapeutic agents had been to devise a test-tube system or laboratory animal model to simulate the disease process or to culture the infectious agent under study. Then many, even hundreds, of chemicals were screened looking for any with potentially therapeutic effect – a procedure that was obviously inefficient and did not make optimal use of professional and other research assets.

In contrast to this trial and error method, the approach pioneered by these workers was to examine and compare the metabolic processes of an infecting microorganism and of normal tissue, and any differences found could then to be exploited as potential targets of attack by chemotherapeutic agents. Using this more rational and efficient methodology, Hitchings and Elion identified and developed three important antimicrobial agents; pyrimethamine, trimethoprim and acyclovir.

Pyrimethamine

Pyrimethamine is a chemotherapeutic agent that interferes with the synthesis of folic acid and the derivatives which are required in turn for the synthesis of nucleic acids (Figure 7.1). The drug

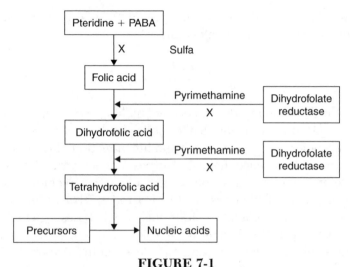

FIGURE 7-1
The mechanism of action of pyrimethamine.

acts by interfering with the enzyme dihydrofolate reductase, which converts folic acid to dihydrofolic acid and then to tetrahydrofolic acid. This enzymatic reaction is required for DNA synthesis both by bacteria and by mammalian cells; however, the dihydrofolate reductase enzyme of bacteria is 100 000 times more sensitive to pyrimethamine than is the mammalian enzyme, and it is this differential sensitivity that accounts for the utility of the drug in treating microbial infections. Pyrimethamine is usually used in combination with a sulfonamide, thus providing inhibition of two sequential steps in folic acid metabolism. Pyrimethamine plus a sulfa is used in the treatment of malaria and toxoplasmosis, a protozoan infection which is of particular importance in causing congenital disease.

Trimethoprim

Trimethoprim, a second compound developed by Hitchings and Elion, is similar to pyrimethamine both in structure and mechanism of action. In combination with sulfamethoxazole it is used to treat bacterial infections of the urinary tract, as well as certain venereal and respiratory infections.

Acyclovir

The third compound developed by Hitchings and Elion is acyclovir, an antiviral agent that inhibits the herpes viruses that cause 'cold sores', eye infection and inflammation of the central nervous system (encephalitis). Herpes simplex virus type 1 is the most common cause of corneal ulceration (keratitis) and corneal cause of blindness in the US today. Structurally, acyclovir is similar to one of the four nucleic acid bases – guanine – that make up DNA. The antiviral acts by inhibiting DNA polymerase, an enzyme produced and utilized both by normal cells and the herpes viruses to synthesize DNA. Acyclovir itself does not have antiviral activity; it first must be taken up and activated by a viral enzyme thymidine kinase. The host cell also has its own thymidine kinase; however, the viral enzyme has 200 times greater affinity than the cellular enzyme for acyclovir. The result of this affinity differential is that acyclovir is taken up and activated mostly in virus-infected rather than in normal host cells. Accordingly, the DNA polymerase of the virus is inhibited but the normal host's enzyme is not.

The treatment of herpes simplex keratitis with acyclovir eye drops in 1977 was the first virus infection managed by a specific chemotherapeutic antiviral agent. Since then a number of derivatives of acyclovir have been developed and are now used in other diseases caused by the human herpes viruses.

Part C

BACTERIA

The classes of organisms that cause infectious diseases are, in ascending order of biological complexity and organization, the viruses, bacteria, fungi, unicellular and multicellular parasites.

Bacteria are ubiquitous microorganisms, occurring in the air, the soil, on the skin and in the gastrointestinal tract. While bacteria are major causes of infectious diseases, the majority of them are benign, harmless and even useful.

In general, all living cells, both plants and animals, are *eucaryotes* – that is, the nucleus with its chromosomal genetic DNA complement is separated from the cytoplasm of the cell by a discrete nuclear membrane. Bacteria, in contrast, constitute the class of *prokaryote*, in which the chromosomal apparatus is not separated from the rest of the cell contents – that is, there is no discrete nucleus.

Bacteria are classified into major groups based upon four basic characteristics:

1. **Morphology**
2. **Staining characteristics**
3. **Spore formation**
4. **Oxygen requirements.**

Morphologically, bacteria can be spherical (*cocci*), elongated and rod-shaped (*bacilli*) or coiled (*spirochetes*).

Bacteria are of about one micron in size – 25 000 per inch – and can be visualized only with a relatively high-powered light microscope. To be visible under the microscope bacteria must first be stained, and the most common technique used is the Gram stain, which was devised in 1884 by Hans Christian Joachim Gram, a Danish microbiologist. In this procedure bacteria are fixed on to a microscope slide, treated with the purple dye 'crystal violet', plus iodine, and then washed with alcohol. Those bacteria that retain the stain after the alcohol wash are termed Gram-positive, whilst those not retaining the stain are termed Gram-negative. The unique importance and utility of this simple staining procedure is not just that it provides ready visualization of the organisms but rather that it correlates with important biological features of the species of bacteria – most notably the types of disease produced and patterns of their susceptibility to the various antibiotics.

Most bacteria can be killed by heating to 150°F for thirty minutes or 161°F for fifteen seconds – the procedure used in pasteurization of milk. However, some important bacteria, notably the clostridia (which cause tetanus, gas gangrene and botulism) produce resistant spores that require steaming under increased pressure for inactivation (250°F for fifteen minutes).

Some bacteria, like cells of higher organisms, require oxygen for growth; these bacteria are the *aerobes*. Others, however, the anerobes, will only grow in the absence of oxygen – a unique characteristic first described by Pasteur.

Bacteria can be cultivated in various liquid media and on media solidified with agar, a polysaccharide derived from seaweed. The unique feature of culture on solid medium is that each individual colony which develops is derived from a single bacterial cell in the inoculum – that is, the colony is a *clone*.

Bacteria multiply by simple binary fission, an asexual process in which a single bacterial cell divides in two. By the application of genetic techniques Joshua Lederberg showed that a sexual process of recombination between two different bacterial cells can occur. This discovery was recognized by a share of the Nobel Prize for Physiology or Medicine in 1958.

Bacteria can divide as rapidly as once every twenty minutes, so that a single cell can initiate seventy-two generations and five million progeny in twenty-four hours. This feature makes bacteria a uniquely convenient laboratory model system for the study of genetics and mutations.

Bacteria produce disease by a variety of mechanisms, including production of exotoxins or endotoxins, destruction of host tissues with abscess formation and, in some cases, inciting in the host a detrimental auto-destruction immune response by the host. Exotoxins are proteins that are secreted into the host environment and which inhibit specific metabolic processes. Examples include diphtheria, tetanus, gas gangrene, food poisoning of various kinds, and various types of dysentery. Other bacteria produce endotoxins, which are components of the bacterial cell wall, and cause shock which is frequently irreversible and lethal. In addition, some bacteria cause harmful immune responses by the host, a prime example being the cardiac and joint injuries of acute rheumatic fever that results from a streptococcal sore throat.

Stephen Jay Gould advanced the thesis that bacteria, not insects, rule the world – a formulation which has much to recommend it; bacteria are the machinery of the biological world.

Table C.1 provides a summary of the classes and biological characteristics of bacteria, and also examples of diseases caused by them.

Table C.1. Bacteria.

Class	Biological characteristics	Examples of diseases
Chlamydia	Obligate intracellular growth	Trachoma (eye disease) Type of pneumonia
Rickettsia	Obligate intracellular growth	Typhus Spotted fevers
Mycoplasma	No cell wall	Type of pneumonia
True bacteria	Aerobic, Gram (+) Aerobic, Gram (−) Anaerobic, Gram (+) Anaerobic, Gram (−)	Abscesses Gonorrhea Tetanus Diphtheria Abscesses
Acid-fast bacteria	Special stain required	Tuberculosis Leprosy
Fungal-like bacteria	Common in soil, growth in chains	Source of many antibiotics

8

TUBERCULOSIS

The captain of all these men of death that came against him to take him away, was the Consumption, for it was that that brought him down to the grave.

(JOHN BUNYAN, 1628–1688)

ROBERT KOCH

The 1905 Nobel Prize for Physiology or Medicine was awarded to Robert Koch for his discovery of the cause of tuberculosis, *Mycobacterium tuberculosis.*

Robert Heinrich Germann Koch (11 December 1843–27 May 1910) was a German physician and a founder of the science of bacteriology. He graduated in medicine in 1866 at the University of Göttingen. He practiced medicine, was a field surgeon during the Franco-Prussian War (1870–1871) and was then district surgeon in Wallstein. It was there that he established a bacteriology laboratory, first for studies of algae and later for pathogenic bacteria. In 1877 he transferred to the German Health Office, again setting up a bacteriology laboratory. He announced his discoveries regarding the tubercle bacillus and tuberculosis in 1882. Subsequently he did field studies in Egypt and in India on plague and cholera, and in Africa on sleeping sickness. During his professional career he made significant contributions, in addition to tuberculosis, to understanding anthrax, cholera, malaria, amebic dysentery and bacterial conjunctivitis.

THE HISTORY OF TUBERCULOSIS

Tuberculosis is truly a disease of antiquity; evidence of spinal infection has been found in pre-Colombian remains in South America, and in Egyptian mummies. Pulmonary tuberculosis was described by Hippocrates, and the disease was known in India from at least 500 BC.

Tuberculosis became a major problem in Europe in the seventeenth and eighteenth centuries, with the advent of the Industrial Revolution. Populations became crowded together in sub-optimal housing and public health conditions, resulting in decreased host resistance to infections. During these times one-quarter of all adult deaths resulted from tuberculosis.

Early in the twentieth century, with improved hygienic and social conditions and then later with the introduction of effective chemotherapy, the incidence of tuberculosis declined. More recently, however, because of the AIDS epidemic, tuberculosis has had a resurgence. Unfortunately multi-drug resistant tubercle bacilli have recently appeared, and the overall result of these trends is that tuberculosis remains the most frequent infectious disease cause of death in the world today.

KOCH'S CONTRIBUTIONS

Koch's outstanding contribution, which established him as the father of medical bacteriology, was his conceptual insights that guided his creation and initiation of the basic procedures necessary for scientific investigations in microbiology. These procedures remain fundamental for studies in medical bacteriology even today. He recognized, first and foremost, that microorganisms occur in nature in mixed populations and that, in order for accurate and meaningful studies, pure cultures containing only a single species must be utilized. To this end he devised the use of solid rather than liquid culture procedures. The nutrient media were solidified with agar-agar and cultures planted on these media. He originally used glass slides to support the solid media but subsequently used shallow dishes with glass lids, an innovation of his laboratory assistant, Julius Richard Petri. Petri dishes remain standard in bacteriology laboratories today.

The key concept involved was the recognition that a single bacterial cell would grow into a macroscopic colony, and that this represented a clone – that is, it derived from a single cell. These clones or pure culture colonies could then be transferred and studied free of contamination by other bacteria.

Koch also pioneered the use of aniline dyes to stain bacteria, which permitted not only their recognition but also species differentiation under the microscope.

Further, Koch proposed four criteria for establishing an organism as the cause of a specific disease. These criteria, the famous 'Koch's Postulates', are:

1. The organism must be found first in the diseased but not in the normal host
2. The organism must first be isolated from the diseased animal and grown in pure culture
3. The pure culture must produce the disease when injected into a normal animal
4. The organism must be re-isolated from the experimentally infected animal.

Using these basic tools, Koch isolated *M. tuberculosis*, the cause of tuberculosis, a discovery he reported in 1882.

The organism initially escaped detection by the usual staining techniques because of its uniquely resistant lipid cell wall. Koch, however, was able to stain the organism by using prolonged immersion of the organism plus heat to promote uptake of the stain. This use of heat, according to popular folklore, was a serendipitous result of his wife's leaving some slides on a heater.

The primary staining procedure for the tubercle bacillus was devised by Paul Ehrlich in 1882. He found that aniline dyes mixed with phenol would stain essentially all bacteria but that, uniquely, the stained tubercle bacillus resisted de-staining with a dilute acid solution. This property, termed 'acid-fast', defines the genus mycobacteria.

Koch also devised a coagulated blood serum medium and inoculation at body temperature for culture of the bacterium. He held these cultures for prolonged periods of time, finding that colonies began to appear only after ten days or two weeks, in contrast to other ordinary bacteria which appeared in twenty-four to forty-eight hours.

Further, Koch devised an extract of tubercle bacillus broth culture that he proposed as a potential treatment for the infection, although this unfortunately did not prove useful. However, this extract, tuberculin, became the basis of a skin test for assessing exposure to the bacillus. A more refined form of Koch's 'Old Tuberculin' purified protein derivative (PPD) continues in use as a skin test reagent today. In this test, a small quantity of PPD is injected into the skin of the forearm. If a person has been exposed to tuberculosis the site of injection will become indurated in forty-eight hours – an immune response that will persist for several days before subsiding. This is a positive tuberculin test.

SPREAD OF TUBERCULOSIS

Tuberculosis is an airborne disease, spread from an infected person by coughing. The infectious aerosol is taken in by the susceptible into the lung, where it is either contained or can further spread to multiple sites in the body, including the central nervous system and bone. The lung infection can lead to cavity formation, and those sufferers with cavities are particularly efficient aerosolizers of the bacterium to others.

Tuberculosis is a highly contagious disease. Examples of its spread include the infection of members of an audience at a rock concert where one of the singers had active pulmonary disease, and an outbreak on a naval vessel in which the secondary cases were all shown to be on the same subcircuit of the ship's ventilation system as the index case.

While it may seem contradictory, considering the worldwide mortality from tuberculosis, recognizable clinical disease caused by the infection is somewhat infrequent. Approximately 80 to 90 percent of individuals who are infected remain asymptomatic for life. They show no symptoms, their chest X-rays are normal or may have only a small scar, and their only evidence of contact with the organism is a positive tuberculin skin test.

SYMPTOMS

Tuberculosis is in most cases a chronic illness, running a course of years with slowly progressive wasting and weight loss – hence the common term 'consumption'.

The anemia of the infection prompted the name 'The White Plague', which ravaged seventeenth and eighteenth century Europe, wiping out one-quarter of its population. The anemia of The White Plague is due both to coughing up blood from diseased lungs and to failure of the bone marrow to produce replacement red blood cells.

The custom of applying white powder to the face is said to have arisen from the desire of courtesans to emulate the appearance of the social and intellectual elite of the day, many of whom had tuberculosis. The list of the eighteenth and nineteenth century elite who were tuberculosis victims reads like a *Who's Who* of the era. They included Frédéric Chopin, Niccolò Paganini, Jean Jacques Rousseau, Anton Chekhov, Friedrich von Schiller, Laurence Sterne, Percy Shelley, Edgar Allan Poe, Eugene O'Neill, Sir Walter Scott, and Robert Louis Stevenson. In addition, Dr Edward Livingston Trudeau, the founder of the tuberculosis sanatorium in Saranac Lake, also suffered from the infection.

TREATMENT

The therapy for tuberculosis in the nineteenth and early twentieth centuries was carried out in sanatoria specifically established for the disease. The only therapy was 'healthy living', which consisted of eating nutritious food and taking exercise.

A major tuberculosis sanatorium was founded in 1884 at Saranac Lake, New York, by Dr Edward Trudeau, who also was a victim of the infection. Features were open-air treatment along with outdoor living and sleeping. The months and years spent in the sanatorium were helpful to patients because it allowed time for their innate host defenses to wall off and contain the infection. In the winter of 1887–1888, Robert Louis Stevenson was treated at Saranac Lake. Another famous sanatorium, at Davos, Switzerland, was the setting for the novel *The Magic Mountain*, written by Thomas Mann in 1924.

The lung cavities of pulmonary tuberculosis always posed the hazard of spontaneous and heavy hemorrhage, and accordingly a good deal of effort was expended in trying to collapse the cavities. Various methods used included pumping air into the abdominal cavity to elevate the diaphragms, and removal of ribs (thoracoplasty) from the affected side in order to facilitate lung collapse.

The entire therapeutic picture changed in the mid-twentieth century. In 1946 streptomycin was found to be effective against the tubercle bacillus (see Chapter 6). In 1952 INH was released for anti-tuberculosis therapy, and in combination with streptomycin plus a third antituberculous drug was found to be effective therapy. PAS (para-amino salicylic acid) was frequently used as the third drug, as it prevented the emergence of drug-resistant bacilli. Because of this effective drug regimen the sanatoria and tuberculosis wards of hospitals were closed and patients were treated on an outpatient basis.

Currently there are approximately one dozen anti-tuberculosis chemotherapeutic drugs available. Despite this array there is an emerging problem of multi-drug resistant tuberculosis cases, particularly in the Far East and in Eastern Europe. These infections are refractory to therapy and in fact pose a significant hazard to the population of the rest of the world. The specter of pre-drug therapy worldwide tuberculosis is both real and sobering.

VACCINATION

A vaccine for prevention of tuberculosis was developed by two students of Pasteur; Albert Calmette and Camille Guerin. The vaccine (BCG) consisted of a live strain of *Mycobacterium tuberculosis* attenuated by passage numerous times on bile-containing medium. The vaccine, which was initially produced for infants and was first used in 1922, is currently used in all age groups. It is widely used throughout the world, but has not been generally accepted in the United States because controlled tests to demonstrate its efficacy have yielded highly variable results, for reasons not entirely clear. It probably has more effect in controlling the disseminated form of the disease seen in infants than in the pulmonary form occurring in adults.

BCG is generally safe, although occasional cases of systemic disease due to the vaccine organism have been reported. Safety has always been a concern, even to the originators of the vaccine, especially because of its use in infants.

An unfortunate error resulted in what is called the 'Lübeck disaster'. In the spring of 1930 in Lübeck, Germany, 249 infants were vaccinated with BCG, and by autumn of that year 73 had died of tuberculosis. A thorough investigation exonerated BCG after it was found that the batch of vaccine used had accidentally been contaminated with virulent *Mycobacterium tuberculosis*. Since that time strains of virulent tubercle bacilli have been proscribed from buildings where BCG vaccine is manufactured.

With the current emergence of multi-drug resistant tuberculosis, there is an increased interest in vaccination as the future hope for ultimately controlling the infection. In addition to standard BCG, other candidate vaccines are being considered – including subunits of the tubercle bacillus; DNA from the tubercle bacillus which upon administration codes for production of tubercle bacillus antigens; and other bacteria genetically altered to produce tubercle bacillus antigens.

Despite all of the advances made by medical science in controlling tuberculosis, it remains the number one cause of infectious disease deaths in the world today.

OTHER MYCOBACTERIA

The genus mycobacterium includes, in addition to the tubercle bacillus, the leprosy bacillus (Hansen's bacillus) and a group of organisms collectively termed 'atypical mycobacteria'.

Leprosy is an ancient disease spread by person-to-person contact. It has always been much feared because of the deformities it produces. Lepers were once social outcasts, abandoned by family and friends. In Europe they were required to wear a bell around their necks to warn others of their presence. They were often confined to isolation in sanatoria in remote areas; the most famous being the Molakai, Hawaii, colony administered by Father Damian. His bacterial infection after 20 years of service was announced by his arresting salutation on a radio broadcast: 'We lepers …'.

The atypical mycobacteria are a group of soil bacteria of low virulence that cause chronic pulmonary infection, usually in individuals whose lungs have been previously damaged by other infections including classical tuberculosis.

9

TYPHUS

Plague follows the trade routes;
Typhus follows the wars.

(ANONYMOUS MEDICAL MAXIM)

If communism does not destroy the louse,
the louse will destroy communism.

(VLADIMER ILICH LENIN, 1870–1924)

CHARLES NICOLLE

The 1928 Nobel Prize for Physiology or Medicine was awarded to Charles Nicolle for his discovery that typhus is transmitted by the louse.

Charles-Jules-Henri Nicolle (21 September 1866–28 February 1936) was a French physician who received his medical degree from the school in Rouen. He practiced there, and was also on the faculty of the medical school. His interest was in medical bacteriology, after having taken a course in the subject at the Pasteur Institute. In 1902 he was appointed Director of the Pasteur Institute in Tunis where, until 1932, he studied typhus as well as other infectious diseases including brucellosis, tuberculosis, measles and diphtheria.

HISTORY OF TYPHUS

Epidemic or classical typhus is an infectious disease of historic as well as medical importance. It was responsible for the plague of Athens (430–428 BC), contributed to Napoleon's defeat in Russia (1812), and compounded the hardships of the Irish potato famine (1845–1850). Typhus was prevalent during both the First and Second World Wars in Europe. In the Second World War DDT was available and was widely used to quell the epidemic (see Chapter 25).

Classical typhus epidemics occur where people are crowded together in unsanitary conditions. These factors account for such synonyms for the disease as 'gaol (jail) fever', 'camp (military and refugee) fever' and 'war fever'.

EPIDEMIC TYPHUS

Classical epidemic typhus is an infectious disease characterized by fever, headache and a prominent rash which starts centrally on the trunk and then spreads centrifugally to the limbs. Untreated, the mortality rate can be as high as 40 percent.

The cause of epidemic typhus is a rickettsia, *Rickettsia prowazekii*. The rickettsiae are small bacteria which will grow only within living cells – that is, they are obligate intracellular parasites (Table 9.1). There are many species of rickettsiae, only some of which can cause disease in man; the organisms are primarily pathogens of animals and insects.

The rickettsia of epidemic typhus was first described by Rocha Lima in 1916. He named the organism *Rickettsia prowazekii* after two pioneer investigators, both of whom had died of the infection.

The transmission of epidemic typhus by lice was first demonstrated by Nicolle in 1909. His discovery was stimulated by the observation that secondary cases in the hospital occurred among personnel who first admitted the patients, but after the patients

Table 9.1.   Rickettsial infections.

Group	Species	Disease	Vector	Reservoir
Typhus	R. prowazekii	Epidemic typhus	Lice	Man
		Brill–Zinsser	None	None
	R. typhi	Endemic typhus	Fleas	Rats
	O. tsutsugamuchi*	Scrub typhus	Mites	Rodents
Spotted	R. rickettsii	Rocky mountain	Ticks	Ticks,
fever		spotted fever		animals
	R. akari	Rickettsialpox	Mites	Mice
	(and others)			
Miscellaneous	C. Burnetii**	Q Fever	?Ticks	Cattle,
				sheep,
				goats

* Orientia, ** Coxiella.

had been bathed and given clean clothing, no subsequent transmission occurred. This observation focused Nicolle's attention on the patients' clothing, and he found that the lice present in their clothes were found to be the critical disease vector. The transmission of classical epidemic typhus by lice was confirmed in 1910 in Mexico by Henry Taylor Ricketts, best known for his work on another rickettsial disease, Rocky Mountain Spotted Fever.

Lice tend to be host-specific, and the louse of humans is *Pediculus hominis*. There are two varieties of human lice; body lice and head lice. Transmission of epidemic typhus is related primarily to body lice. The lice live on the skin and also in the clothing. The adult lice lay eggs, termed *nits*, which develop into nymphs, go through three molts and emerge as adults. Both the nymphs and the adults are bloodsucking insects.

The transmission of rickettsia begins when the louse takes a blood meal from an infected individual. The organisms multiply in the gastrointestinal tract of the louse and are passed in the insect feces. The louse bite causes itching, and when the uninfected host scratches the bite the organisms in the infected louse feces are introduced. The rickettsial infection ultimately kills the louse in about two weeks.

In some individuals the clinical manifestations of epidemic typhus become quiescent but the organisms remain dormant in the body. After some years, possibly because of advancing age or declining immunity, the disease again becomes active. This has been observed in some Eastern Europeans years after their emigration to the US. Reactivation epidemic typhus is referred to as 'Brill–Zinsser disease', named after the two investigators who described the syndrome in 1934. The inter-epidemic reservoir of *Rickettsia prowazekii* has not been identified, but possibly the Brill–Zinsser carrier of the infection may serve this function. In addition there have been suggestions that there may possibly be unrecognized animal reservoirs in nature, the most exotic of these candidates being the flying squirrel.

ENDEMIC TYPHUS

During his studies in Mexico, Ricketts found that not all cases of typhus seemed to be transmitted by lice – an observation that had previously been made by Nicolle. These cases were milder than the classical epidemic disease and untreated were seldom fatal. This disease, termed *endemic typhus*, was shown to be due to a distinct rickettsial species, *Rickettsia typhi*, which is transmitted from rats to humans by the rat flea (see Table 9.1).

10

SYPHILIS THERAPY

It was true indeed that I had got the sickness; but
I believe I caught it from that fine young servant-girl whom
I was keeping when my house was robbed. The French
disease, for it was that, remained in me more than four
months dormant before it showed itself, and then it broke
out over my whole body at one instant. It was not like
what one commonly observes, but covered my flesh with
certain blisters, of the size of six-pence, and rose-
colored. The doctors would not call it the French disease,
albeit I told them why I thought it was that.

I went on treating myself according to their {the doctors'}
methods, but derived no benefit. At last, then, I resolved
on taking the wood {guaiac}, against the advice of the
first physicians in Rome; and I took it with the most
scrupulous discipline and rules of abstinence that could
be thought of; and after a few days, I perceived in me a
great amendment. The result was that at the end of fifty
days I was cured and as sound as a fish in water.

(BENVENUTO CELLINI, 1500–1571)

JULIUS WAGNER-JAUREGG

Julius Wagner-Jauregg was awarded the 1927 Nobel Prize for
Physiology or Medicine for his use of medically-induced malaria
infection to treat neurosyphilis.

Julius Wagner Ritter von Jauregg (7 March 1857–27 September 1940) was an Austrian psychiatrist and neurologist who obtained his medical education at the University of Vienna. He was a member of the psychiatric staff at the University from 1883 to 1889, and Professor of Psychiatry at the University of Graz from 1889 to 1893. From 1893 to 1928 he was Professor of Psychiatry and Director of the Hospital for Nervous and Mental Disorders, Graz.

MALARIOTHERAPY

While at the University of Vienna, Wagner-Jauregg observed that patients with certain mental disorders seemed to improve if they contracted an intercurrent febrile infection. This improvement with fever was particularly striking, and potentially useful, in those patients whose underlying neurological disease was syphilis involving the brain.

He first attempted to treat neurosyphilis patients by inoculating them with streptococci, the bacterial cause of erysipelas, a systemic infection characterized by fever and generalized rash. This was discontinued, however, because of serious consequences of the infection. He next attempted to induce fever with injection of tuberculin (see Chapter 8), but found this procedure largely ineffective. In 1917, while at the University of Vienna, he inoculated neurosyphilis patients with the blood of malaria cases and found this to produce beneficial effects. An advantage of this procedure was that quinine could be used, if necessary, to control the fever if the malaria infection was excessively severe. After further studies he concluded that malariotherapy plus neosalvarsan, a chemotherapeutic agent, gave the best results. On one occasion he accidentally inoculated four patients with *Plasmodium falciparum* malaria, which produces severe disease, rather than the intended milder species, *Plasmodium vivax*, and this resulted in three deaths.

Workers in other centers treated neurosyphilis patients with malariotherapy and found that in about one-half of the patients there was either complete remission or marked improvement. Fever therapy was effective because the causative agent of syphilis, *Treponema pallidum*, is heat-labile and is inhibited by the elevated body temperature. After malaria-induced fever therapy was shown to be effective, fever cabinets – steam cabinets in which the patients sat – were introduced, accomplishing the same result as the malaria infection.

Action

The fever of infection is in fact an adaptive protective mechanism by the host rather than a pathogenic mechanism of the microorganism. Elevation of temperature increases the activity of the host's defense mechanism as well as directly inhibiting the infecting agent. This is the underlying mechanism of malariotherapy.

In infections in general there has always been some difference of medical opinion as to how fever should be managed. Some elevation of temperature is helpful; however, fever makes the patient uncomfortable. Aspirin or similar drugs can then be given to moderate the temperature elevation.

There are, however, occasional instances where the temperature rises excessively, to 105°F or above, when there is resulting cell death. These instances, termed 'hyperpyrexia', constitute medical emergencies, and temperature must be lowered quickly with such measures as immersion in cold water, and use of ice packs and alcohol sponges.

HISTORY OF SYPHILIS

The origins of syphilis continue to be a matter of historical controversy. The Old World (Pre-Colombian) theory holds that the disease originated in Central Africa and was carried to Europe by traders and travelers. The New World (Colombian) theory is that the disease was endemic in Haiti and was transported by Columbus's crew back to Spain after his second expedition to the New World in 1492–1493. By 1497 the infection was widespread in Europe; in France it was known as the Naples disease, in England as the Spanish disease, and in Naples as the French

disease. By the beginning of the twentieth century it was estimated, for both Europe and the United States, that the lifetime
risk of acquiring syphilis was between 5 percent and 20 percent.
At this time neurosyphilis was responsible for 20 percent of hos-
pitalizations on psychiatric services.

It was not until 1905 that two German workers in Berlin, Fritz
Schaudin and Eric Hoffman, observed the spirochete and named
it *Spirochaeta pallida* – the genus name because of the spiral or
corkscrew shape of the bacterium, and the species name pallida
because it did not take up the usual bacterial stain (i.e. it was
pale). The current nomenclature is *Treponema pallidum*.

CLINICAL SYPHILIS

Clinically, syphilis is divided into three stages; primary, secondary and tertiary.

Primary syphilis occurs with sexual exposure, and the characteristic local lesion or *chancre*, usually on the genitalia, contains the
infectious organisms. From this lesion the treponema dissemi-
nate throughout the body, while the primary lesion heals spontaneously in a few weeks.

Following an interval of several months secondary syphilis occurs.
This is characterized by a generalized skin rash, mucous membrane lesions, lymph node enlargement and fever. The lesions of
primary and secondary syphilis are highly infectious. Again, as
with primary syphilis, these signs and symptoms regress without
treatment, although there can be relapses. Following secondary
syphilis there is a period of latency, which may last for years, during which there are no observable signs or symptoms of disease.

Latent syphilis does not progress in two-thirds of the patients;
however, in the remaining one-third tertiary syphilis super-
venes. There are three forms of tertiary syphilis:

**1. Late benign syphilis is a chronic disorder in which
lesions, called *gummas*, can occur anywhere in or on the body,
tending to progress by erosion of underlying tissue or bone.**

2. **Cardiovascular syphilis is a chronic disorder, with involvement of the main artery carrying blood from the heart to the body (the aorta).**
3. **Neurosyphilis appears in three different clinical forms. The first is meningovascular syphilis, which can present as an acute to subacute meningitis or even as a stroke. The second is tabes dorsalis, in which the patient experiences difficulty in walking and maintaining balance. The third is general paresis, which is the form that the psychiatrist sees; this was the major target for Wagner-Jauregg's therapeutic studies. These patients are usually demented and have lost all contact with reality. They may claim to be Napoleon or the King of England, or have other grandiose delusions.**

TREATMENT

Prior to fever therapy the only available treatments for syphilis were injections of calomel (mercury chloride), and mercury-containing ointments. In 1910 Paul Ehrlich introduced an organic arsenic compound, salvarsan (arsphenamine), which was usually administered by injection in combination with either bismuth or mercury. However, patients so treated still progressed to neurosyphilis, and for this reason fever therapy was a significant advance in the treatment of syphilis.

Fever therapy for syphilis was replaced following the discovery of penicillin, which remains the treatment of choice today. With the advent of this antibiotic, research on *Treponema pallidum* essentially ceased; in particular, the search for a vaccine is no longer considered a priority. However, syphilis has by no means been eradicated. While there was a progressive decline in the number of reported cases over the decades with penicillin treatment, there has more recently been a resurgence concomitant with the worldwide AIDS epidemic (see Counterpoint, at the end of this book).

What is old is new.

Part D
VIRUSES

Viruses are obligate intracellular parasites – that is, they replicate only within living cells, not in the extracellular environment of the host or in cell-free artificial media (Table D.1).

Viruses infect animals and plants. Bacteria are also subject to viruses termed 'bacteriophage'.

The size of viruses, depending upon the species, can range from 0.02 to 0.2 microns; thus they are approximately between one-fiftieth and one-fifth of the size of typical bacteria.

Morphologically, plant and animal viruses are either spherical, or elongated cylinders. The spherical form is actually an icosahedron, a regular twenty-sided structure. The cylindrical viruses are constructed of a tubular sheath of protein containing the nucleic acid core. The morphology of bacteriophage is more complex; they are composed of a spherical 'head' with an elongated 'tail'.

The functional anatomy of all viruses is the same; a nucleic acid core surrounded by a protein coat. The viral nucleic acid carries the genetic information for self-replication, while the protein coat allows the virus to attach to the host cell and also protects the viral genome from host enzymes that break down nucleic acids.

In all plant and animal cells the genetic information is encoded in genes composed of DNA, which are strung together on chromosomes constituting the cell's genome. The genes code for RNA *transcription*, which in turn directs protein synthesis, *translation*. Viruses, however, are the exception to this rule; their genome consists of either DNA or RNA but never both, with either nucleic acid serving as the genome.

Table D.1.   The human herpes viruses.

Designation	Common name	Disease
HHV-1	Herpes Simplex Virus type 1	'Cold sores' Keratitis Encephalitis
HHV-2	Herpes Simplex Virus type 2	Genital herpes
HHV-3	Varicella-Zoster virus	Chickenpox Shingles
HHV-4	Epstein–Barr Virus	Infectious mononucleosis Burkitt's lymphoma Nasopharyngeal carcinoma
HHV-5	Cytomegalovirus	Congenital infections Retinitis Intestinal infections in transplants
HHV-6		Exanthem subitum (Roseola infantum)
HHV-7		Exanthum subitum-like illness
HHV-8	Kaposi's sarcoma- associated virus	Kaposi's sarcoma

To begin the process of replication, viruses first attach to and enter the host cell. Animal viruses attach by specific receptors and are then taken into the cell by a process termed endocytosis. Plant viruses invade through areas of physical damage or abrasion to the plant surface. Bacteriophage use a more complex strategy. They attach to the host cell by their 'tail' and inject their nucleic acid genome, like a needle and syringe, the protein coat remaining outside. This finding provided one of the original pieces of evidence that genetic information is carried by nucleic acid and not, as previously thought, by protein.

Viruses do not possess the complex biochemical machinery necessary for self-replication; rather they commandeer and subvert the host cells' machinery for their own reproduction.

Viral infection of a cell can have one of three outcomes:

1. *Lysis*. In the lytic (or productive) cycle, virus enters the cell, is uncoated, and the viral nucleic acid codes for the replication of

more nucleic acid and for coat protein. These components are then assembled into new viral progeny and released to infect additional cells.

   2. *Temporate infection*. In the temperate infection, the viral nucleic acid is integrated into the host cell genome and multiplies in concert with the host cell; no new complete virus is formed or released.

   3. *Tumor production*. Tumor production by viruses has been recognized for almost as long as viruses themselves. Tumor virology has now expanded to the point where it is a valid and extensive study in itself. The attraction of this field is multifaceted; it not only affords a method for investigating the basic biology of tumors, but also presents the intriguing possibility that vaccine to prevent or antiviral agents to treat could be developed.

Defense against virus infection is provided by the immune system (see Part A). In the extracellular environment viruses are attached by antibody, while replicating intracellularly the immune system, via suppressor cytotoxic T-cells, destroys the infected host cells. In addition, interferon is an important system with activity against many RNA and some DNA virus infections. The term 'interferon' was coined in 1957 by the discoverers Isaacs and Lindemann. They observed that a substance was produced in influenza A-infected chick embryos which, upon transfer, protected normal chick embryos against viral infection. It is now recognized that the function of interferons is not limited to the originally described antiviral activity. Interferons also inhibit host cell protein synthesis, stimulate phagocyte cells, and inhibit the growth of certain tumor cells. The full range of their activities has probably not yet been defined. Currently interferons are being used clinically to treat a variety of diseases, both viral and nonviral, such as multiple sclerosis, hepatitis B and hepatitis C.

Viruses are not classified as cells; they are a simpler life form. While they are the most primitive of biological agents, they were probably not the initial form of life to appear on Earth as they require host cells for growth. Viruses possibly arose originally from the host cell genome, which became enclosed in protein and became transmissible cell-to-cell. As viruses are dependent upon the host cell for replication, it is not surprising that the simplest host cell, the bacteria, replicates the most complex of viruses, the bacteriophage, while the most complex of host cells, mammalian, replicates the simplest viruses, such as poliovirus.

Viruses are universally accepted as being living agents on the basis that they enter the host cell and direct their self-replication. It is interesting to note that virus can be stored in freezers for years, at which time they are no more than complexes of organic chemicals, like other chemicals in a laboratory. 'Suspended animation'?

11

TOBACCO MOSAIC VIRUS

It is an experiment, as all life is an experiment.

(OLIVER WENDELL HOMES, JR, 1841–1935)

WENDELL STANLEY

Wendell Stanley received a share of the 1946 Nobel Prize for Chemistry for his crystallization of tobacco mosaic virus (TMV).

Wendell Meredith Stanley (16 August 1904–15 June 1971) was an American biochemist. He obtained his PhD from the University of Illinois, and then pursued post-doctoral studies at the University of Munich, Germany. From 1932 to 1948 he worked at the Rockefeller Institute for Medical Research at Princeton, New Jersey. From 1948 to 1971 he was Professor of Biochemistry and Director of the Laboratory for Virus Research at the University of California, Berkeley. During the Second World War he developed a vaccine for viral influenza.

THE TOBACCO MOSAIC VIRUS

Tobacco mosaic virus causes a mottled browning of tobacco leaves, and accordingly is of major economic importance. It also infects other crops, most notably tomatoes. The virus is spread mechanically from infected plants to scratched or damaged leaves of normal plants. The virus is very stable in nature, and has been isolated from tobacco produce several years after their preparation. The only available agricultural control measure is to destroy infected plants.

Tobacco mosaic virus played a central role in establishing the field of virology; it was in fact the first virus to be recognized. In 1879, Adolph Mayer, director of the Agricultural Experiment Station in Wageningen, Holland, undertook studies on the diseases of tobacco. He showed that a leaf-mottling disease could be transmitted by rubbing juice from diseased plants onto the leaves of healthy plants. He coined the name 'tobacco mosaic disease' and suggested that the etiology was bacterial, although no such agent could be cultured.

The discovery of viruses is attributed to Dmitry Ivanovsky, a Russian microbiologist who, between 1887 and 1890, investigated the mosaic disease of tobacco plants occurring in Eastern Europe. He found that the disease-causing factor passed through a porcelain filter that had pores fine enough to hold back bacteria. He thus demonstrated that the cause of tobacco disease was due to a 'filterable virus' – the term *virus* coming from the Latin meaning *poison*. This term is now considered obsolete and is abbreviated to simply 'virus'.

Ivanovsky's work was confirmed in 1895 by the Dutch botanist Willem Beijerinck, who in addition showed that the disease produced by the filtered sap could be transferred serially plant-to-plant, proving that the agent was living and self-replicating rather than being simply a chemical toxin.

Vinson and Petri, in a series of studies at the Boyce Thompson Institute, Philadelphia, between 1927 and 1931, concentrated the virus by precipitation using the established laboratory methods developed for proteins.

In 1935, while at the Rockefeller Institute, Stanley purified tobacco mosaic virus and derived fine needle-like crystalline preparations which were fully active and infectious. This achievement was not only the initial crystallization of a virus, but also the unprecedented and unexpected example of a living entity occurring in the crystalline state.

Tobacco mosaic virus (TMV) was subsequently shown to be a single-stranded RNA virus of filamentous morphology.

OTHER PLANT-INFECTING VIRUSES

Tobacco mosaic virus is by no means the only virus infecting plants. There are now recognized additional plant viruses such as tomato spotted wilt virus, sugar cane mosaic virus, citrus quick decline virus, barley yellow dwarf virus, etc. These viruses cause severe crop losses and are of major economic importance worldwide.

Not all plant viruses are harmful, however, either to the plants or to the horticulturist. An outstanding example is the tulip breaking virus. Tulips are native to wide regions of Eurasia and were first introduce into Antwerp, Holland, from Constantinople (Istanbul) in 1562. The plants rapidly became popular, first with professional gardeners and then with speculators. In the Netherlands, the period between 1633 and 1637 became known as the 'Tulip Mania' or 'Tulip Craze'. The flowers occur either in solid colors, termed 'self-colored', or are streaked in variegated colors, termed 'broken', and the latter varieties became much sought after, with rapidly escalating prices. It is said that at this time a single rare bulb could sell for as much as three times the price of a Rembrandt portrait. Prices skyrocketed, almost like a Ponzi scheme, until suddenly in 1637 doubt arose that the price rise could be maintained and the market collapsed. It is now recognized that the broken variation in tulips is the result of a viral infection.

STANLEY'S ACCOMPLISHMENT

Stanley's specific accomplishment was the crystallization of tobacco mosaic virus; however, its biological significance far transcended its chemical triumph. The demonstration that a living entity could be crystallized, like an ordinary laboratory-shelf chemical such as table salt or sugar, challenged the accepted distinction between what is living and what is not.

There were two noteworthy experimental laboratory findings which represented major steps that preceded the conceptual consequence of Stanley's work. The first was Friedrich Wöhler's 1828 synthesis of urea, an organic compound which was made from inorganic chemicals. This achievement overturned the prevailing dogma at the time, the 'Theory of Vitalism', which held that there was a vital force in organic material that placed them beyond chemical laboratory synthesis. Organic material, it was believed, could only be synthesized by living entities. The second was the S. F. Miller–H. C. Urey experiment of 1953, which showed that if an electric discharge was passed through a closed vessel containing water, hydrogen, methane and ammonia, then amino acids and hydroxy acids were produced. This laboratory setup was designed to mimic conditions that existed on the primordial earth, with the electric discharge being equivalent to lightning, suggesting the origin of life on earth.

Stanley's work showed that a crystalline chemical could also be a living entity. The prions (see Chapter 22) could be the next challenge to blur the classical distinction between living and non-living entities.

12

YELLOW FEVER

**Humanity has but three great enemies;
Fever, famine and war; of these by far the greatest,
by far the most terrible, is fever.**

(SIR WILLIAM OSLER, 1849–1919)

MAX THEILER

The 1951 Nobel Prize for Physiology or Medicine was awarded to Max Theiler for his development of a vaccine for the prevention of yellow fever.

Max Theiler (30 January 1899–11 August 1972) was a South African-born American microbiologist. He attended the University of Capetown, and in 1920 traveled to St Thomas' Hospital and the London School of Hygiene and Tropical Medicine for medical training. In 1922 he joined the Department of Tropical Medicine at Harvard Medical School, Boston. From 1930 to 1964 he continued research at the Rockefeller Institute for Medical Research (the Rockefeller University) in New York City, where he developed the yellow fever vaccine.

YELLOW FEVER

Yellow fever occurs in man and in a variety of monkeys and small jungle mammals, and is transmitted by the mosquito. It is characterized by jaundice (hence the name) resulting from liver involvement, and hemorrhage from the gastrointestinal tract. The mortality rate can be up to 50 percent, and there is no specific antiviral therapy available.

Two interrelated epidemiological patterns involve infections in man. In sylvatic yellow fever the infection is maintained by several species of mosquitoes among monkeys and other mammals, and man is incidentally infected upon entry into that environment. In urban yellow fever the disease is propagated among city dwellers by the anopheline mosquito.

History

Yellow fever has been recognized since the seventeenth century. It is endemic in West Africa and in central and coastal South America, and outbreaks have also occurred in the eastern United States and in western Europe. In the 1870s a catastrophic outbreak occurred in Memphis, Tennessee, killing 5000 residents and crippling the city's economy for ten to fifteen years. The last major US outbreak was in New Orleans, Louisiana in 1905 resulting in 5000 cases and 1000 deaths. Currently there are an estimated 200 000 cases annually worldwide.

Yellow fever was the first virus isolated from disease in man. This was accomplished by Walter Reed, in Cuba in 1901, who showed that the bacteria-free serum of yellow fever victims could transmit the infection.

The demonstration that yellow fever is mosquito transmitted provides one of the classic stories in epidemiology. This mode of transmission was first suggested in 1881 by Carlos Finley, a Cuban epidemiologist. This was contrary to the belief held at that time that transmission was by fomites – that is, by contact with personal effects of victims, such as clothing or bedding. In 1900, Walter Reed, a US Army pathologist and bacteriologist, went to Cuba as head of a commission sent to investigate an outbreak, originally thought to be malaria but subsequently found to be yellow fever. To pursue these studies, James Caroll, an Army surgeon in Havana, was exposed to a mosquito that had fed on a yellow fever patient, resulting in a typical clinical case. Jesse Lazear, a physician and commission member, was inadvertently bitten by a yellow fever-infected mosquito and died of the resulting infection.

To document the role of the mosquito in transmission, a small camp was set up with volunteer subjects. One group was housed in a barracks that was supplied with unwashed linen and clothing from yellow fever patients but was protected from mosquitoes by screening on windows and doorways. The second group was housed in quarters supplied with fresh linen and clothing but without mosquito-proof screening. Yellow fever occurred only in the barracks without screens. This experiment discredited the fomite theory and established the mosquito vector mode of transmission of the disease.

The Army Corps of Engineers, under the leadership of Major W. C. Gorgas, was sent to eliminate the mosquitoes from Havana, and in three months the area was yellow fever-free.

TREATMENT AND CONTROL

Max Theiler's contribution was the development of a live avirulent strain of yellow fever virus for use as a vaccine. His first vaccine consisted of virus that had been attenuated by passage in monkeys, and this was used by the French in Africa. He then developed a safer vaccine that had been passed in chick embryos, and this preparation (vaccine strain 17D) is currently used. A single dose of this live virus vaccine produces long-lasting immunity, although current public health regulations require a repeat dose at ten-year intervals.

A number of procedures are in use in endemic areas to control the mosquito vectors. These include eliminating the sources of stagnant water where the mosquitoes breed and the larvae develop. Water can also be treated with temephor (Abate), and organophosphorous insecticide sprays may be used. Minnows of Gambusia fish, which feed on mosquito larvae, are stocked in swampy water areas as an environmentally friendly control measure. More recently, male mosquitoes rendered sterile by irradiation are being released to decrease breeding efficiency.

However, the possibility of worldwide eradication of any infectious disease depends upon its epidemiology and the presence of reservoirs in nature. In the case of yellow fever, the animal reservoirs in nature make global eradication very unlikely. Public health measures, including mosquito eradication and vaccination of susceptibles, will continue to be necessary indefinitely to control the infection.

In contrast, where there is no reservoir in nature, such as with smallpox and poliomyelitis, eradication is possible. This has already been accomplished with the former and is being attempted with the latter. For this reason smallpox vaccination has until just recently been discontinued and polio vaccination has become a worldwide priority.

The current problem with smallpox is that stocks of the virus are maintained in both the United States and in Russia. In addition to these known stocks, it is also conceivable that there may be other stocks of smallpox virus occult at the present time. The threat of biological warfare using this virus has emerged and has prompted selective vaccination programs, especially among the military and healthcare workers. Although statistically rare, complications resulting from smallpox vaccination, including generalized vaccinia, progressive vaccinia and cardiomyopathies, are already occurring or must be anticipated. Not surprisingly, the cost:benefit ratio of this current vaccination program is being debated.

13

POLIOMYELITIS VIRUS

Some folk seem glad even to draw their breath.

(WILLIAM MORRIS, 1834–1896)

JOHN ENDERS, THOMAS WELLER AND FREDERICK ROBBINS

John Enders, Thomas Weller and Frederick Robbins shared the 1954 Nobel Prize for Physiology or Medicine for cultivating poliovirus *in vitro* in human embryonic skin and muscle tissue culture.

John Franklin Enders (10 February 1897–8 September 1985) was an American virologist and microbiologist who received his PhD degree from Harvard University. His early studies were on tuberculosis and pneumonia, and on resistance to infectious diseases. His research on viral infections was carried out at the Children's Hospital in Boston, where he contributed to the *in vitro* culture methods for both polio virus and measles virus.

Thomas Huckle Weller (15 June 1915–) is an American virologist who earned his MD degree from Harvard University in 1940. His studies were carried out in Enders' infectious disease laboratory at the Children's Hospital in Boston. He contributed to the tissue culture of poliovirus, and of rubella (German measles) and varicella (chickenpox) viruses. He became Professor of Tropical Public Health, and from 1966 to 1981 was Director of the Center for Prevention of Infectious Disease at the Harvard School of Public Health.

Frederick Chapman Robbins (25 August 1916–4 August 2002) was an American pediatrician and laboratory researcher who earned his MD degree at Harvard University. His work on the cultivation of poliovirus was done at the Children's Hospital, Boston. He then moved to Case Western Reserve University in Cleveland, Ohio, as Professor of Pediatrics and then as Dean of the Medical School in 1966. In 1980 he became President of the Institute of Medicine at the National Academy of Sciences. In 1985 he returned to Case Western Reserve, and in 1990 he was appointed Chairman of the International Commission for the Certification of Poliomyelitis Eradication in The Americas.

POLIOMYELITIS

Prior to the work of Enders, Weller and Robbins, poliovirus could only be studied in monkeys – making any critical studies expensive, cumbersome and time-consuming. Their tissue culture method revolutionized studies on poliovirus as well as on many other viruses. In this procedure embryonic skin and muscle tissue from aborted fetuses was collected, treated with trypsin to free individual cells, and then grown in a nutrient medium in tubes or bottles. Poliovirus was then cultured in these cells and could be harvested in large quantities for chemical analysis and immunologic analyses, as well as for production of vaccines.

Symptoms and spread

Poliomyelitis is one of the epidemic infectious diseases of man that has been recognized since ancient times. It has also been known as 'infantile paralysis' because of its propensity to affect the very young. The name 'poliomyelitis' refers to the gray matter (nerve cells) of the central nervous system, and myelitis, specifically the spinal cord.

The disease is characterized by fever and muscle paralysis which can affect part of a limb, an entire limb, or all four limbs. Nerve involvement can affect the muscles of the chest, making breathing difficult or impossible. The paralysis which results from poliovirus infection runs a variable course from complete resolution to permanent disability. A particularly severe form of paralysis, termed bulbar polio, affects the centers at the top of the spinal cord and results in the inability to swallow food or water or saliva.

Polio virus occurs naturally only in man; there is no animal reservoir in nature. The virus infection initially invades the gastrointestinal tract and in most cases is eliminated from this site, without further spread, by the normal immune defenses. In 1 per cent to 2 per cent of cases it may, however, involve the local lymphatics and then infect the motor nerves of the spinal cord, with resulting paralysis.

The importance of the fecal–oral route of transmission was shown by the early observation that polio cases in children clustered around public swimming pools. For this reason children were often kept away from these popular summer recreation centers.

The important defensive role of the lymph nodes is shown by the association between tonsillectomy and bulbar polio – the tonsils being lymph nodes in the throat.

Management

The management of the established poliovirus infection is directed primarily toward the paralyses produced – a problem that is now confined almost exclusively to the developing regions of the world. There are no antiviral agents for treatment of polio infection. Paralysis of the limbs is treated with passive movement physical therapy, a method championed by Sister Elizabeth Kenny in the 1950s.

Paralysis of the respiratory muscles requires assisted ventilation. This was originally performed by the 'iron lung', the tank ventilator invented by Dreiber and Shaw in the 1920s and manufactured by the Emerson Company of Boston. This mechanical device completely enclosed the body, leaving only the patient's head exposed. The interior of the chamber was cycled between vacuum and normal atmospheric pressure every four to five seconds, resulting in respiratory chest movement. In the 1930s and 1940s hospitals had wards with dozens of these machines for managing affected children. The device was expensive, cumbersome, uncomfortable, and impeded routine nursing care of the patient. Currently there are more sophisticated respiratory assist machines which interface with the patient via only an endotracheal tube.

Vaccination

There are three different serotypes of poliovirus. Each produces lifelong immunity that is type-specific and does not extend to the other two types. Accordingly, a vaccine must contain all three serotypes.

The current control of polio is by means of vaccines; the Salk inactivated polio vaccine (IPV) which was introduced in 1955, and the Sabin live oral polio virus vaccine (OPV) which was introduced in 1962.

The IPV is composed of the three polio serotypes, I, II and III, grown in continuous cell lines derived from monkey kidney tissue. The virus is then killed with formalin to produce the vaccine, which is given as three intramuscular or subcutaneous doses at eight-week intervals beginning at age two months, followed by a booster at age four to six years. The advantage of the IPV is that it is safe, containing no live virus. Its disadvantages are that it requires multiple injections over a period of time and so is less well suited for use in the third world, and that IPV also does not prevent gastrointestinal infection with the virus; it only prevents spread from the gastrointestinal tract to the nervous system.

The immunized person can therefore still have an infection and pass the virus on without showing typical paralytic poliomyelitis – thus the IPV protects the individual but not the community. The new enhanced-potency IPV now used, however, does produce some gastrointestinal immunity, but less than OPV.

The OPV consists of the three serotypes which have been attenuated by passage in monkey kidney tissue culture, following which strains are selected that are not neurovirulent for monkeys. These strains are then grown in monkey kidney tissue culture, purified and prepared as a live attenuated trivalent vaccine for oral administration. The advantage of the OPV is that it can be given as a single dose – which is convenient, technically simple and well suited for use in the third world – and, in contrast to the IPV, the OPV also immunizes the gastrointestinal tract and thus is useful for preventing person-to-person spread of the infection. The major disadvantage is that it is a live virus vaccine, and so it must be kept refrigerated and also cannot be used in individuals with immune deficiencies.

Initially Salk IPV was the only vaccine available, and accordingly was widely used. When the Sabin OPV was introduced in 1962 this new single-dose vaccine became the predominant vaccine in use. Since 1979 there have been only about eight cases of polio annually in the US. All of the cases were shown to be due to the virus being in the OPV itself, and for this reason the enhanced potency IPV is now used in the US and has replaced the OPV.

Since 1996 polio has been eradicated from the western hemisphere, but the infection remains in the Indian subcontinent and in sub-Saharan Africa. In polio-free areas, unless vaccination is continued the current population will be replaced by polio susceptibles who can be infected from cases in the remaining endemic areas. The eventual worldwide eradication of the infection is possible; however, if immunization is discontinued, the infection, like smallpox, could be reintroduced maliciously as a weapon of bioterrorism. Polio would not be a prime candidate for this misuse, though, as the majority of infected individuals do not become clinically ill, and permanent paralysis is unusual.

The importance of the tissue culture method of Enders, Weller and Robbins is not confined to the control solely of poliomyelitis. It has also supplied the vaccines for control of other epidemic infectious diseases such as rubeola (red measles), rubella (German measles), mumps and varicella (chickenpox). Further, their methodology has provided the standard laboratory method for study, diagnosis and control of many of the known virus infections of man.

14

HEPATITIS B VIRUS

Our ideas are only intellectual instruments which we use
to break into phenomena; we must change them when
they have served their purpose, as we change a blunt
lancet that we have used long enough.

(CLAUDE BERNARD, 1813–1878)

BARUCH BLUMBERG

Baruch Blumberg received a share of the 1976 Nobel Prize for
Physiology or Medicine for his discovery of the hepatitis B virus.

Baruch S. Blumberg (28 July 1925–), who was born in New York
City, received his MD degree at Columbia University and his PhD
in biochemistry at Oxford University in 1957. He then engaged
in research at the National Institutes of Health, Bethesda, Maryland, following which he was appointed to the University of
Pennsylvania and the Institute for Cancer Research, Philadelphia.
His basic interest was the mechanisms by which different groups
and populations varied in their susceptibilities and responses to
diverse diseases.

HEPATITIS

Hepatitis designates inflammation of the liver, and is caused
by a number of different infectious agents. It is characterized

by jaundice, which is a yellowish discoloration of the skin resulting from a failure of inflamed liver cells to excrete bile pigments.

One form of hepatitis, which has been known for centuries, is spread from person to person and can occur in major epidemics. A second form of hepatitis was first seen in 1937 in a group of Brennan harbor shipyard workers who had all received a yellow fever vaccine that had been stabilized by the addition of human serum. The disease transmitted person to person in epidemics was termed *infectious hepatitis* or hepatitis A, while the disease transmitted by parenteral inoculation of human serum preparation is termed *homologous serum jaundice* or hepatitis B. This distinction was confirmed by epidemiological studies in the 1960s at the Willowbrook State School for the Mentally Handicapped in New York, an institution where, essentially, all children entering soon developed hepatitis. These studies were later criticized on the basis that the studies should not have been carried out – rather, the institution should have been closed down.

In Blumberg's studies at the National Institutes of Health beginning in the early 1960s, blood samples were examined for variations in their protein composition in an effort to study world population interrelationships and their migrations. The reagents used to detect these variations were sera from hemophiliacs, because they routinely receive repeated blood transfusions and accordingly develop antibodies to unusual blood components. A novel protein component was detected in the blood of an Australian aborigine using the serum from an American hemophiliac, and this protein component was termed *Australia antigen*. The Australia antigen was subsequently found in the sera of patients with such diverse disorders as leukemia, leprosy and Down's syndrome (mongolism), and also in some normals. By the late 1960s it became apparent that the Australia antigen was found specifically in the serum of patients with homologous serum jaundice (hepatitis B) and was in fact an antigenic component of the hepatitis B virus (HBV), the cause of the disease.

Hepatitis B

Hepatitis B is a common infection in the general population, with an incidence of between three and four per cent. In renal dialysis patients, who are continuously exposed to blood transfusions, up to 40 per cent may be infected.

Hepatitis B is transmitted by injection of blood products derived from an infected person. It occurs commonly among intravenous drug users who share the same non-sterile hypodermic needles. It is also spread via sexual contact. An important mode of transmission is from mother to offspring at the time of birth, or via breastfeeding.

The major health risk of hepatitis B is not the acute disease; rather it is the subsequent development of chronic liver disease, which occurs in about 20 per cent of those infected. The chronic infections progress to scarring of the liver, *cirrhosis*, and then, in some, to liver cancer (hepatocellular carcinoma).

The diagnosis of hepatitis B depends upon measurement of specific antibody in the serum and the detection of virus or viral products in the serum. For the past fifteen years units of blood for donation have been screened for hepatitis B so that the risk of infection is in the order of one case per 60 000 units of transfused blood.

The control of hepatitis B is based upon a vaccine, specific immune serum, and the antiviral agent lamiduvine. The source of the hepatitis B vaccine was originally the serum of infected patients, from which the virus was concentrated, purified and killed. There was always concern about this source because of the possibility that unidentified and undetected viruses of some other type could be present and survive the preparatory procedure. To obviate this risk, and also to increase production efficacy, the current vaccine is prepared from yeast cell 'factories' in which hepatitis B genes are added to the yeast cell chromosomes, which then produce hepatitis B antigens.

A unique feature of hepatitis B virus is that the neonatally infected have a 90 per cent risk of developing a chronic persistent infection. For this reason universal immunization of the new-born is now the accepted standard of care. In addition, hepatitis B vaccination is now routine in healthcare workers, in those exposed to blood or blood products, and in those traveling to developing countries of the world.

Hepatitis B infection is usually self-limiting and does not cause chronic disease. However, it can persist leading to scarring of the liver (cirrhosis). When cirrhosis does occur then hepatocellular carcinoma can follow. Chronic infection leading to cirrhosis occurs in about 1 per cent of hepatitis B virus-infected adults, and in these adults the risk of hepatocellular carcinoma is about 3 per cent per year.

If a person is exposed to hepatitis B, prevention can be accomplished with an injection of gamma globulin with has a high titer anti-hepatitis B antibody followed by vaccination. An important group for prevention care is the newborn of mothers having the virus circulating in the bloodstream. These infants are given hepatitis B immune serum at birth and started on the course of vaccination.

The genome of hepatitis B virus is a partially single-stranded and partially double-stranded DNA. Upon infection, the single-stranded segment must be duplicated by an enzyme that has reverse transcriptase activity. For this reason it is inhibited by the anti-reverse transcriptase drug lamiduvine.

With the widespread use of hepatitis B vaccine, the incidence of chronic liver impairment and liver cancer resulting from the virus infection has significantly decreased. Treatment of hepatitis B is with the antiviral agent, lamivudine, which is also used for HIV/AIDS (see Counterpoint at the end of this book).

Hepatitis A

Hepatitis A occurs primarily in children, and produces a mild and self-limiting disease. It is transmitted by fecal–oral contact, and is readily spread by direct contact – especially in schools and daycare centers.

Two recent food-borne local epidemics linked to raw green onions have occurred. Among the adults infected several deaths have occurred – an outcome practically never seen in children. This age-related general pattern is repeated in other epidemic infectious diseases. With improvements in sanitation, and other preventive measures, the occurrence of infections previously common in children is delayed until adulthood, when the consequences can be much more severe. Currently there is killed virus vaccine prepared in human fibroblast tissue cultures for prevention of hepatitis A.

Hepatitis C

After hepatitis A and B had been identified it became apparent that there were additional cases of hepatitis not associated with these two agents, and these were referred to as non-A non-B hepatitis. These cases, like hepatitis B, seemed to be associated with blood transfusions, illicit drug use and sexual activity. In the 1980s this virus, now termed hepatitis C virus (HCV), was identified by molecular biological techniques from the blood of infected patients. Like hepatitis B it can be either mild and self-limiting, or can progress to chronic liver disease, cirrhosis, and liver cancer. Blood for transfusion is now screened to exclude the presence of hepatitis C. Persons already infected can now be treated with a combination of interferon, a biological immune product, and ribavirin, an antiviral agent.

Approximately 3 per cent of the world population is infected – 170 million people – with hepatitis C. Antibody tests for infection, as well as tests directly for the hepatitis C virus RNA, are now routinely done on all blood donated for transfusion so that the risks of contracting hepatitis C from this source are minimal.

Illicit drug use is a major means of transmission of HCV; sexual transmission, however, seems to be of only minor importance.

The major health risks of hepatitis C virus are the persistence of the virus infection in the liver, which occurs in 70 per cent of cases and after a latent period of twenty to thirty years, with cirrhosis followed by the risk of hepatocellular carcinoma. With better control of hepatitis B, hepatitis C is now the most common viral cause of hepatocellular carcinoma.

As hepatitis C virus differs from hepatitis B virus in that its genome does not integrate into the cellular genome, another mechanism is involved in the production of hepatocellular carcinoma. In the case of hepatitis C virus, proteins that it produces which have been shown to protect liver cells from programmed cell death (apoptosis) may be the responsible factors.

Currently, chronic hepatitis C infections are being treated with interferon-α plus ribovirin in an attempt to prevent cirrhosis and hepatocellular carcinoma.

DNA viruses also are associated with cancer, although the relationship between the infection and tumors is not as well defined as with the RNA viruses.

Other forms of hepatitis

Another hepatitis virus, hepatitis D, has been identified, primarily in Italy and northern Mediterranean Europe. It is a virus

Table 14.1.   Basic characteristics of the five currently recognized hepatitis viruses.

Hepatitis virus	Genome	Incubation (days)	Epidemiology	Prevention	Treatment
A	RNA	15–45	Fecal–oral	Immunoglobulin	None
B	DNA	30–180	Injection, Perinatal, Sexual	Specific immunoglobulin vaccine	Interferon, lamivudine
C	RNA	15–160	Injection	None	Interferon, ribaviron
D	RNA	30–180	Injection, Perinatal, Sexual	Hepatitis B vaccine*	Interferon
E	RNA	14–60	Fecal–Oral	None	None

*Because hepatitis D only occurs with simultaneous hepatitis B, the hepatitis B vaccine is indicated.

with a bizarre ecology, occurring only in patients already infected with hepatitis B virus. There is a great deal more to be learned about the basic biology of this unique two-virus interrelationship. Such a relationship is, however, not unknown – for example, it occurs between adenovirus and adeno-associated virus.

A fifth hepatitis virus, hepatitis E, has recently been identified in Mexico. In most features it resembles hepatitis A, producing an acute self-limiting infection.

The spectrum of known hepatitis viruses has expanded significantly, although it seems that the major actors have now appeared on stage (Table 14.1). The history of infectious disease, however, suggests that this might not be an entirely safe wager.

15

BACTERIOPHAGE

So, naturalists observe, a flea
Hath smaller fleas that on him prey;
And these have smaller still to bite 'em;
And so proceed *ad infinitum*.

(JONATHAN SWIFT, 1667–1745)

ALFRED HERSHEY

Alfred D. Hershey received a share of the 1969 Nobel Prize for Physiology and Medicine for his studies on the biology and genetics of bacteriophage infection.

Alfred Day Hershey (4 December 1908–22 May 1997) was an American microbiologist who received his PhD in 1934 from Michigan State University, East Lansing, and was a professor at Washington University, St Louis, from 1934 to 1950. He then joined the Genetics Research Unit of the Carnegie Institution of Washington, and carried out research at Cold Spring Harbor Laboratories, New York.

BACTERIOPHAGE

Bacteriophage, or bacterial viruses, were first recognized by F. W. Twort in 1915 as a degenerative change in cultures of staphylococcus isolated from calf lymph. He found that this degeneration could be passed serially to fresh culture of the bacteria. In 1917, Fi d'Herelle independently observed the same phenomenon in mixed cultures of intestinal bacteria and showed that the active agent could pass through filters that held back bacteria. This 'Twort–d'Herelle' phenomenon was due to a virus that attacked bacteria. It is now recognized that there are bacteriophage that infect essentially every known species of bacteria.

Bacteriophage are known that have either DNA or RNA as their genetic material, in either circular or linear configuration, and as a single- or a double-stranded molecule. The morphology of various bacteriophage can be diverse, but most commonly they consist of a 'head' containing the nucleic acid and a 'tail' by which the virus attaches itself to the susceptible bacteria and through which its nucleic acid is injected to initiate the infection.

HERSHEY'S CONTRIBUTION

Hershey was a member of the 'phage group', an informal network of physicists-turned-biologists who studied and exchanged their experimental results on bacteriophage. The success of this group was, in part at least, attributable to the fact that they all worked with the same system – the T series of bacteriophage which infect the common colon bacterium *Escherichia coli*.

In 1945 Hershey showed that spontaneous mutations occurred in both bacteriophage and in bacterial hosts, similar to those previously known to occur only in higher life forms. In 1946 he demonstrated that upon simultaneous infection of a single bacterial cell with two different bacteriophage, reassortment of their DNA could occur. This again is similar to the recombination that occurs between the two chromosomes of a pair in higher forms.

In 1952, Hershey, with Martha Chase, investigated the mechanism by which the genetic information of the bacteriophage is transferred to the host bacterial cell. In this now famous

Hershey–Chase experiment, bacteriophage-infected bacteria were grown in a medium containing radioactive sulfur and radioactive phosphorus. This yielded a stock of bacteriophage in which their nucleic acid contained radioactive phosphorus and their protein contained radioactive sulfur.

These differentially labeled bacteriophage were used to infect a culture of bacteria. After allowing a few minutes for bacteriophage–bacteria absorption, the bacteriophage were sheared off by agitation in a Waring blender. When the removed bacteriophage and the bacteria were tested separately, the removed bacteriophage contained only radioactive sulfur (indicating protein), while the bacteria now contained the radioactive phosphorus (indicating nucleic acid). The bacteria were then further incubated, and analysis of the new bacteriophage produced showed that their nucleic acid contained the phosphorus radioactive label but the protein did not contain the radioactive sulfur label. This experiment showed that only the nucleic acid of the bacteriophage entered the bacterial cell, and thus the DNA carried all of the information necessary for the synthesis of new complete bacteriophage.

Prior to this experiment it was debated whether the genetic information was coded for by DNA or by protein. In fact protein, which was composed of twenty different amino acid components, seemed a more likely candidate than DNA, which was composed of only four different bases. The Hershey–Chase experiment established that the genetic information was carried by DNA. It is upon this foundation, plus the Watson–Crick elucidation of the three-dimensional structure and triplet code of DNA, that the current molecular biology era is constructed.

Clearly the next revolution of equal magnitude will be the understanding of how the brain acquires, stores, integrates and retrieves information. Brain – the 'Final Frontier'.

16

BACTERIOPHAGE LYSOGENY

The shrewd guess, the fertile hypothesis, the courageous leap to a tentative conclusion – these are the most valuable coin of the thinker at work.

(JEROME SEYMOUR BRUNER, 1915–)

ANDRÉ LWOFF

André Lwoff received a share of the 1965 Nobel Prize for Physiology or Medicine for his studies on the interaction of bacteriophage, the viruses that infect bacteria, with their host cells.

André-Michael Lwoff (8 May 1902–30 September 1994) was a French biologist who attended the University of Paris. His research career was primarily at the Pasteur Institute, Paris, where he was on the board of directors from 1966 to 1972. He was also Professor of Microbiology at the Sorbonne, Paris, and from 1968 to 1972 was Director of the Cancer Research Institute at Villejuif. He was awarded the Medal of Resistance and made an officer in the Legion of Honour for his efforts during the Second World War, in the French underground.

BACTERIOPHAGE INFECTION

There are two types of bacteriophage infection of bacteria; these are termed *virulent* and *temperate* (*lysogenic*).

In the virulent infection the bacteriophage injects its nucleic acid into the bacteria and, within a matter of just a few hours, several hundred new bacteriophage particles are produced. The bacterial cell is then lysed, releasing the new bacteriophage particles, which are available to infect additional bacterial cells.

The second type of infection, discovered by Lwoff in 1950, is a temperate interaction of bacteriophage with the host bacterial cell. In this type of infection the bacteriophage nucleic acid integrates into the bacterial genome and is replicated in synchrony with the bacterial chromosome, generation after generation, without harm to the bacterium. Lwoff termed this type of infection *lysogeny*.

One novel consequence of lysogeny is that bacteria carrying the bacteriophage can acquire metabolic processes not present in the uninfected state. This is termed *lysogenic conversion*. An example of lysogenic conversion occurs with the diphtheria bacillus, *Corynebacterium diphtheriae*. The diphtheria exotoxin is produced only by lysogenized organisms, not by the uninfected microorganism. This process of lysogeny and lysogenic conversion occurs with other toxin-producing microorganisms such as *Clostridium tetani*. Further, this general process of adding genetic information to a cell's chromosome, which is then transcribed and translated into a product, is the basis of the current commercial production of viral vaccines in yeast cells. It is also the biological basis for the genetically modified food crops that are gratefully welcomed by some and reviled by others.

In addition, there may be genetic information present and maintained in the chromosomes of animals that may have originally been of viral origin. It is estimated that up to one-quarter of the human genome has no known function; the origin may have been lysogenic-type viral infections that occurred in the remote past – simply junk genes in the chromosomal attic.

17

ROUS SARCOMA VIRUS

Whenever a new discovery is reported to the
scientific world, they say first, "It is probably not true."
Thereafter, when the truth of the new proposition has
been demonstrated beyond question, they say,
"Yes it may be true, but it is not important."
Finally, when sufficient time has elapsed to fully
evidence its importance, they say, "Yes, surely it is
important, but it is no longer new."

(MICHEL DE MONTAIGNE, 1533–1592)

PEYTON ROUS

Peyton Rous shared the 1966 Nobel Prize for Physiology or Medicine for his discovery that sarcomas in chickens are caused by a filterable agent.

Francis Peyton Rous (5 October 1879–16 February 1970) was born in Maryland. In 1905 he received his MD degree from Johns Hopkins University. After additional studies at the University of Michigan he joined the staff of the Rockefeller Institute in 1909, where he remained for his entire professional career.

His Nobel Prize-winning work was done in 1910, and his receipt of the award in 1966 set two records; the longest interim between discovery and award (fifty-six years), and the designation of the oldest laureate (eighty-seven years). Rous remarked during his Nobel Prize acceptance address that he felt very fortunate still to be alive to receive the award.

FOWL SARCOMA

Rous investigated fowl sarcoma at the behest of a local poultry breeder. First Rous demonstrated that the sarcoma could be transferred by grafting into normal birds. He then prepared a cell-free extract of the sarcoma by emulsifying the tissue and passing it through a filter that held back bacteria, and found that this cell-free extract also produced the sarcoma.

The Rous sarcoma virus contains an RNA genome. In addition to this virus, several other RNA viruses have been found to produce tumors in animals, mammary tumors and leukemias in mice, leukemias in cats, and lymphomas in cattle.

Tumor production by these RNA viruses is the result of their gene, termed an *oncogene*, which over-stimulates the normal host mechanism that promotes cell proliferation.

Rous's suggestion that the fowl sarcoma was caused by a virus was initially rejected by the scientific community but was finally accepted years later – hence the protracted lag period for the Nobel recognition.

18

POLYOMA VIRUS

All things are hidden, obscure and debatable if the cause of the phenomena be unknown, but everything is clear if this cause be known.

(LOUIS PASTEUR, 1822–1895)

RENATO DULBECCO

Renato Dulbecco, an Italian virologist, was a co-recipient of the 1975 Nobel Prize for Physiology or Medicine for his studies on the mechanism of tumor production by DNA viruses.

Renato Dulbecco (22 February 1914–) received his MD degree at the University of Turin in 1936, and was a faculty member there for several years. He emigrated to the United States in 1947 and studied viruses, first at Indiana University and then at the California Institute of Technology. From 1963 to 1967 he was a fellow at the Salk Institute for Biological Studies in La Jolla, California. He was then appointed Director of the Imperial Cancer Research Fund in London. He returned to the Salk Institute from 1977 to 1981, and was co-appointed to the faculty of the School of Medicine, University of California at San Diego.

POLYOMA VIRUSES

In order to understand how viruses produce tumors, Dulbecco studied polyoma virus in the test tube rather than in laboratory animals, using transformation as a marker. When normal cells grow in tissue culture they produce a single layer adherent to the glass surface and proliferate until the individual cells just touch each other, at which stage growth stops. Under the influence of tumor viruses this contact inhibition is lost and the cells continue growing, piling up on top of each other. He found that the polyoma viral DNA integrated into the host cell chromosome, and from this site prevented the cell from responding to the tumor suppressor gene products produced by the normal cell. The result is tumor production.

The polyoma viruses plus the related papilloma viruses constitute a family which has as its characteristics a single-stranded DNA genome and the capacity to produce tumors (see Table 18.1). This family was originally termed *papova virus*, a compound term made up of the first two letters of the names of the three species; papilloma virus (*pa*), polyoma virus (*po*) and simian virus 40 (vacuolating agent, *va*).

There are a number of polyoma viruses each of which infect a specific animal species, such as mice, simian species, hamsters, birds and man, producing a wide array of different tumors.

Simian virus 40 (SV_{40}), now classified as a polyoma virus, was originally detected in rhesus monkey kidney tissue culture. This

Table 18.1.   The two groups within the papova family of DNA viruses which cause infections and tumors both in animals and in man.

Group	Virus	Disease
Polyoma virus	Mouse polyoma	Multiple tumors
	Simian virus 40 (SV_{40})	Kidney infection
	Human BK	Kidney infection
	Human JC	Brain infection
Papilloma virus	Animal papillomas	Warts
	Human papillomas	Cancer of uterine cervix, anal and genital tracts, tonsils

virus causes vacuolation of African Green monkey kidney cells grown in tissue culture – hence its original designation 'vacuolating agent'. This virus produces tumors when injected into newborn hamsters. In the 1960s live SV_{40} virus was found to contaminate some lots of killed Salk-type polio virus vaccine. Fortunately, subsequent follow-up of vaccine recipients did not reveal any increased incidence of neoplasia in them.

In addition to the polyoma viruses that infect animals, two members of this group have been found to infect man. These are BK virus, which has been isolated from the urine of a renal transplant recipient, and JC virus, isolated from the brain of an immunosuppressed Hodgkin's Disease patient who had progressive multifocal encephalopathy, a fatal disease of the brain. (The virus designations are from the initials of the original source patients.)

The other group of viruses in the papova family is the papilloma viruses. These viruses are somewhat larger in physical and in genome size than the polyoma viruses. They produce tumors in many different animals, including birds, fish and man. Each virus is host species-specific; that is, cross-infections between animal species do not occur.

Papilloma viruses also infect man, and over 100 types based upon their DNA structure have been identified. Human papilloma virus types 16 and 18 are associated with cancer of the uterine cervix. Other types cause benign skin warts.

The fundamental lesson learned from the study of the polyoma virus is that the DNA viruses cause tumors by integrating into the host cell genome and from this site inhibiting the cellular gene, *tumor suppressive gene*, that normally inhibits cell proliferation. This is in contrast to the RNA tumor viruses, which have a gene (the *oncogene*) that over-stimulates the normal cellular proliferation-promoting mechanism.

19

REVERSE
TRANSCRIPTASE

Your descendants shall gather your fruits.

(VIRGIL, *PUBLIUS VERGILIUS MARO*, 70–19 BC)

HOWARD MARTIN TEMIN AND
DAVID BALTIMORE

Howard Martin Temin and David Baltimore, both American virologists, shared the 1975 Nobel Prize for Physiology or Medicine for their co-discovery of the viral enzyme, reverse transcriptase.

Howard Martin Temin (10 December 1934–9 February 1994) was born in Philadelphia and obtained his PhD under Renato Dulbecco at the California Institute of Technology (Cal Tech). While a graduate student, he studied tumor production by the Rous sarcoma virus. After completing his studies at Cal Tech he joined the faculty at the University of Wisconsin, where he remained throughout his professional career.

David Baltimore (7 March 1938–) was born in New York City, received his doctorate in virology at Rockefeller University in

1964, and did postdoctoral studies at the Massachusetts Institute of Technology (MIT) in Boston. From 1965 to 1968 he studied poliovirus at the Salk Institute in La Jolla, California. He returned to MIT in 1968, and in 1983 he became Director of the Whitehead Institute for Biomedical Research, Cambridge. From 1990 to 1994 he was President of the Rockefeller University, following which he again returned to MIT.

REVERSE TRANSCRIPTION

The studies of Dulbecco on polyoma (see Chapter 18) showed that DNA viruses cause tumors by integrating into the host chromosomal DNA. With the RNA tumor viruses there was no plausible biochemical mechanism by which the viral RNA could integrate into the host DNA. Temin suggested that the RNA tumor viruses code for the synthesis of a DNA intermediate, which then integrates with the host DNA. This suggestion was received with some skepticism, as it contradicted the 'central dogma of molecular biology' – that information flowed from DNA to RNA and then to protein.

Baltimore was at this time studying the RNA tumor virus, Rous sarcoma virus, analyzing the viral enzyme that uses RNA to code for additional RNA. During this investigation at MIT in 1968 he discovered an enzyme that transcribed RNA to produce a DNA complement, thus providing direct evidence for Temin's hypothesis. Transcription is the process by which a DNA template codes for an RNA complement; the enzyme Baltimore discovered was therefore a reverse transcriptase. It is because of the role of reverse transcriptase in RNA tumor virus replication that they are now termed *retroviruses*.

The retroviruses as a family are comprised of a number of agents that infect both man and animals (Table 19.1).

The alpharetroviruses cause leukemias and sarcomas in birds. The deltaretroviruses cause leukemias in man, primates and other animals. One of these, the human T-cell leukemia/lymphoma

Table 19.1.   The retroviruses of man and animals.

Genus	Virus
Alpharetroviruses	Avian leukosis virus (AVL) Rous sarcoma virus (RSV)
Deltaretroviruses	Human T-lymphotropic virus (HTLV)-1, -2 Bovine leukemia virus (BLV) Simian T-lymphotropic virus (STLV)-1, -2, -3
Lentiviruses	Human immunodeficiency virus type 1 (HIV-1) Human immunodeficiency virus type 2 (HIV-2) Simian immunodeficiency virus (SIV) Equine infectious anemia virus (EIAV) Feline immunodeficiency (FIV) Caprine arthritis encephalitis virus (CAEV) Visna/maedi virus

virus, occurs in the Caribbean region and in southern Japan, where it is prevalent but often asymptomatic. Only less than 1 percent of those infected will develop the leukemia/lymphoma malignancy, and then only after a latent period of ten to twenty years.

The lentiviruses infect a range of animals, including man, and characteristically attack and depress the immune system. Of particular importance is the Human Immunodeficiency Virus 1 (HIV-1), which is the cause of the current devastating worldwide pandemic of AIDS.

Because the reverse transcriptase enzyme is present only in the virus and not in man, it is an ideal target for antiviral chemotherapeutic agents. This enzyme opportunity was exploited in the first drugs used to treat HIV/AIDS, and these drugs continue in use today (see Counterpoint, at the end of this book).

20
VIRAL ONCOGENES

We have met the enemy and he is us.

(POGO, BY WALT KELLY, 1913–1973)

MICHAEL BISHOP AND HAROLD VARMUS

The 1989 Nobel Prize for Physiology or Medicine was shared by Michael Bishop and Harold Varmus for their demonstration that viral oncogenes were originally derived from host cell proto-oncogenes.

John Michael Bishop (22 February 1936–) is an American virologist who received his MD degree from Harvard University in 1962. Between 1964 and 1968 he did research in virology at the National Institutes of Health in Bethesda, following which he joined the faculty of the University of California Medical Center, San Francisco. He was promoted to full professorship in 1972, and from 1981 he also served as Director of the University of California George F. Hooper Research Foundation.

Harold E. Varmus (18 December 1939–) is an American virologist who received his MD degree from Columbia University, New York in 1966. After graduation he studied bacteria at the National Institutes of Health, Bethesda. In 1970 he began post-doctoral studies at the University of California, San Francisco, where he collaborated with Bishop. He there became Professor of Biochemistry and Biophysics in 1982. From 1993 to 2003 he was Director of the National Institutes of Health, Bethesda. Currently he is President of the Memorial Sloan–Kettering Cancer Center, New York.

VIRAL ONCOGENES

By the 1970s, the understanding of tumor production by viruses was that the DNA viruses integrate into the host cell genome and interfere with the host cell's tumor suppressor genes or gene products, while the RNA tumor viruses carry their own oncogene via the reverse transcriptase, integrate into the host cell genome, and over-stimulate the cells' proliferation apparatus.

Bishop and Varmus set about to study the oncogenes of the RNA tumor virus. During the course of these studies they found, unexpectedly, that homologues of all the tumor virus's oncogenes could be found in normal uninfected host cells, suggesting that the viral oncogenes were actually host genes that had been commandeered in prior eons from normal host cells. The normal host oncogenes were termed *proto-oncogenes*.

Tumors are an overgrowth of cells which can be induced by a wide range of factors. These factors include chemicals such as certain pesticides, dyes, tobacco products, irradiation such as X-rays and radon gas, and viruses (both DNA- and RNA-containing). Regardless of the specific cause, all operate through a single final pathway; the genetic control mechanism that regulates the appropriate cell population. This genetic mechanism is a two-component reciprocal system; one component, the oncogenes, stimulates the proliferation of cells, while the other, the tumor suppressor genes, inhibits this proliferation. Under circumstances when either the first component is rendered overactive or the second is suppressed, the result is tumor formation.

It is this formulation which has redirected the research on cancer causation and treatment. No longer is the search for viruses that cause cancer; rather the primary efforts are directed towards understanding the genetics and molecular biological mechanisms controlling cell regulation and proliferation. Nonetheless, it was the insights gained by the studies of tumor viruses that revealed this new and productive research arena.

21

KURU

As long as our brain is a mystery, the universe, the reflection of the structure of the brain, will also be a mystery.

(SANTIAGO RAMÓN Y CAJAL, 1852–1934)

CARLETON GAJDUSEK

Carleton Gajdusek shared the 1976 Nobel Prize for Physiology or Medicine for his studies on the epidemiology and transmission of the neurologic disorder, kuru.

Daniel Carleton Gajdusek (9 September 1923–) was born in Yonkers, New York, received an MD degree from Harvard University, and continued with postgraduate studies in pediatrics and infectious diseases there (1949 to 1952). He carried on research at the Walter Reed Army Institute of Research in Washington, DC, the Institute Pasteur in Tehran, and the Walter and Eliza Hall Institute of Medical Research in Melbourne, Australia. In 1958 he was appointed Head of the Laboratory for Virological and Neurological Research at the National Institutes of Health, Bethesda, Maryland.

KURU

While in the South Pacific area in 1955, Gajdusek initiated studies on kuru. Kuru was initially recognized in the Fore tribe in

New Guinea in 1901–1902. The incidence progressively increased to epidemic proportions, and by the mid-1950s kuru was killing 10 per cent of the tribe. In 1957 Zigas and Gajdusek published 'Clinical study of a new syndrome resembling paralysis agitans in Papua natives of the Eastern Highlands of Australian New Guinea'. (Paralysis agitans is a synonym for Parkinson's Disease.)

Kuru means 'trembling' in the Fore language. The disease occurred primarily in children and women; not in adult males. It was characterized by a slow and gradual onset of headache, difficulty with balance and walking, involuntary jerking movements, muscle twitching and weakness, and loss of mental capacity, and terminated fatally in three months to two years after onset. Microscopic examination of kuru victims' brains show degeneration of the tissue leaving tiny holes or vacuoles as in a sponge, hence the synonym *spongiform encephalopathy*. Transmission of the disease was traced to ritual cannibalism, in which the brains of the deceased were rubbed over the bodies of related women and children.

When ritual cannibalism was recognized as transmitting the disease and its practice was outlawed the disease began to die out, and by the 1990s it had virtually disappeared.

In 1965, Gajdusek, Gibbs and Alpers showed that extract of the brains of kuru victims transferred the disease when injected into the brains of chimpanzees, with a latent period of eighteen to thirty-six months. They suggested that kuru was due to slow virus.

KURU-LIKE DISEASES IN ANIMALS

Kuru-like diseases occur in animals as well as in man.

Scrapie

In 1954, B. Sigurdsson, a veterinarian, was studying a form of chronic encephalitis in Icelandic sheep. The disease is characterized by intense itching with resulting loss of wool, altered gait, debility and trembling. The intense itching and rubbing is the source of the name of the illness – 'scrapie'. The incubation period of the disease is approximately two to five years. The infectious agent is present in nasal discharge, and is transmitted via the oral route. The brain shows degeneration of nerve cells with vacuolization, the typical spongiform encephalopathy. The infection can also be transmitted experimentally to mice and hamsters by injecting homogenates of infected sheep brain intracerebally. Sheep scrapie can also be transferred experimentally to Rocky Mountain Elk. Scrapie is the prototype for all of the transmissible spongiform encephalopathies.

Bovine spongiform encephalopathy

Bovine spongiform encephalopathy (BSE, 'mad cow disease') was first recognized in 1988 in the UK, and also occurred in Europe and in Canada. Mad cow disease presents with behavioral changes, agitation, nervousness, lack of coordination, involuntary muscle contractions, weight loss and eventually death. It occurred as a result of feeding ruminant carcass preparations to cattle as a dietary supplement. While this had been done previously, the disease occurred because requirements for processing (including heating or solvent treatment of the supplement) were relaxed. This source of cattle food was banned in 1988. To eliminate this disease, thousands of head of cattle in Great Britain and Europe were slaughtered and their carcasses buried or burned.

In May 2003 a cow in Alberta, Canada, was found to have BSE, and in response the US immediately banned imports of Canadian beef and cattle. The disease was believed to be due to the use of bone marrow containing cattle food possibly prepared from scrapie sheep. No additional cases were found, and by the end of the year the import restrictions were eased to permit importation limited to boneless meat products.

On 23 December 2003, the USDA identified BSE in one six-and-a-half-year-old 'downer' Holstein dairy cow in the state of Washington that had been slaughtered on 9 December. This diagnosis was confirmed by the BSE International Reference Laboratory in Weybridge, England. The cow was traced to its birthplace in Alberta, Canada, and every effort was made to locate and destroy all meat from this cow.

Currently in the US, incorporation of brain and spinal cord into human food is not allowed; nor is the slaughter for human food of 'downer' cattle – that is, those that are too sick to walk.

In the US, only about 120 000 of 35 million cattle slaughtered annually are tested for BSE, and this with a complex chemical test which takes two weeks to complete. In contrast, in Europe all cattle slaughtered that are older than thirty months are tested, and in Japan all slaughtered cattle are tested. Further, Europe and Japan use a test which requires only four to six hours to yield an result, and the USDA is screening applications from companies to provide a comparable rapid test.

Chronic wasting disease of elk and deer

Chronic wasting disease of elk and deer is another transmissible spongiform encephalopathy of animals. This disease was found in Wisconsin in a herd of thirty-seven white tail deer during the 2002 fall hunting season. For this reason the deer herds were submitted to the state for examination during the 2003 hunting season, and again the disease was detected. The mode of transmission among deer is unknown; however, it is believed not to be a hazard to man. In 2002 an elk imported into Korea from Canada in 1997 was found to have chronic wasting disease. Chronic wasting disease was also found in Colorado mule deer.

KURU-LIKE DISEASES IN MAN

It is now recognized that transmissible spongiform encephalopathies other than kuru also occur in man, including Creuztfeldt–Jakob disease (CJD), Gerstmann–Strausseler–Scheinker syndrome (GSS), and fatal familial insomnia (FFI) (see Table 21.1). These

Table 21.1. Transmissible spongiform encephalopathies.

Host	Disease
Man	Kuru Creutzfeldt–Jakob disease Gerstmann–Straussler–Scheinker syndrome Fatal familial insomnia
Animals	Scrapie Bovine spongiform encephalopathy Transmissible mink encephalopathy Chronic wasting disease of elk and deer Feline spongiform encephalopathy Exotic ungulate encephalopathy

Table 21.2. Clinical types of Creutzfeldt–Jakob disease.

Type	Occurrence
s CJD	Sporadic
f CJD	Familial
i CJD	Latrogenic
v CJD	Variant*

* The form of CJD that occurs in man due to eating beef from cattle with bovine spongiform encephalopathy, 'mad cow disease'.

diseases have in common slow onset, inexorable progression to death in months to years, and spongiform changes in the brain as seen by the light microscope. Extracts of brain tissue from these patients can transmit disease to susceptible experimental animals.

Creutzfeldt–Jakob disease was described in the 1920s by two German neurologists, Hans Gerhard Creutzfeldt and Alfons Maria Jakob. CJD presents with memory loss, visual hallucinations, delusions, emotional lability, jerking motions of the limbs and face (myoclonus) and 'startle myoclonus'. CJD can be classified into four categories (see Table 21.2). Most commonly the disease occurs sporadically (s CJD), that is, randomly in the population for no apparent reason. In 10 per cent of cases, however, there is a family history of the disease (f CJD). The disease also can result from medical intervention, including neurosurgical instrumentation, corneal or dural grafts, and the use of human pituitary gland extracts to treat disorders of glandular failure sources of the disease. In 1969 Gibbs and Gadjusek demonstrated that extracts of brain tissue from CJD patients could transmit the disease to primates, indicating that it is one of the transmissible spongiform encephalopathies.

In mid-1996 in the UK and Europe a disease appeared which was diagnosed as a new variant CJD (v CJD). The disorder was characterized by sensory symptoms and psychiatric manifestations. The disease occurred in the young, with an average age of twenty-nine years, and ran a course from onset to death of fourteen months. This is in contrast to classical CJD, which occurs at an average age of sixty-five years and runs a course of four-and-a-half months. In 2002 there were more than 100 deaths from v CJD. The disease is linked to eating beef from cattle suffering from bovine spongiform encephalopathy, 'mad cow disease'.

SUMMARY

The investigations of Gadjusek and colleagues have opened a window onto a whole new class of transmissible neurologic diseases affecting both animals and man. This broad new insight is a remarkable harvest from what appeared initially to be simply an inquiry into an arcane disorder affecting a small native tribe in the South Pacific.

22

PRIONS

When you have eliminated the impossible, whatever remains, _however improbable_, must be the truth.

(SIR ARTHUR CONAN DOYLE, 1859–1930)

STANLEY PRUSINER

In 1997 Stanley Prusiner received the Nobel Prize for Physiology or Medicine for his studies on the nature of the agents that cause the transmissible spongiform encephalopathies; the prions.

Stanley Ben Prusiner (28 May 1942–) was born in Des Moines, Iowa. He received his MD degree from the University of Pennsylvania in 1968, and is currently at the University of California, San Francisco, where he is Professor of Neurology and of Biochemistry and Biophysics at the School of Medicine; Professor of Virology, School of Public Health; and Investigator, Howard Hughes Medical Institute. In 1994 he received the Albert Lasker Award for his studies that subsequently merited the Nobel Prize.

PRIONS

The basic studies on the etiology of the transmissible spongiform encephalopathies were done with scrapie, a chronic, progressive and uniformly fatal disease of sheep and goats characterized by loss of appetite, generalized itching (scrapie) and disturbance in gait and balance. In 1934 this disorder was shown to be transmissible to normal sheep using filtrates of diseased brains injected intracerebrally. In 1961 scrapie was found to be transmissible to mice, also thus providing a convenient laboratory animal model for studying the disease and its causative agent.

The biochemical characteristics of the scrapie agent were extensively examined by Prusiner and reported beginning in 1982. The presence of protein in the agent was shown by its sensitivity to proteolytic enzymes; however, no nucleic acid, either DNA or RNA, could be found based upon the agent's resistance to nucleic acid-destroying enzymes and ultraviolet irradiation. These findings led to the formulation that the scrapie agent is a protein that is devoid of nucleic acid – the 'protein-only hypothesis'. Prusiner designated the protein 'prion' (pronounced *pree-on*) for *pro*teinaceous *in*fectious particle. Because of its filterability and transmissibility, the scrapie agent was initially assumed to be a virus. However, the absence of detectable nucleic acid cast doubt upon this conclusion.

In 1992 a major conceptual advance was reported by Prusiner, who cloned the gene for the scrapie prion and found that this gene was present in the chromosome of the normal host cells. This remarkable discovery revealed that the scrapie prion was coded for by the host cell genome, and was a component of the normal host cell. The major difference between normal prion protein (PrP^C) and the scrapie prion protein (PrP^{SC}) is in their three-dimensional folding configuration. The normal prion protein (PrP^C) molecule is in the α-helical form, which is a spiral spring-like coil, while the abnormal prion protein (PrP^{SC}) is in the β-pleated form, a flattened sheet-like structure.

The normal prion protein (PrP^C) is a surface component of brain cells that functions to protect against oxidative stress and damage. This function is lost upon conversion by refolding to the abnormal prion protein (PrP^{SC}) form.

Scrapie is transmissible, animal-to-animal, because the abnormal scrapie protein (PrP^{SC}) upon contact with the normal protein

(PrP^C) causes its conversion to the abnormal configuration. This conversion is termed *nucleation-growth* or, more colloquially, the 'kiss of death'.

There is variation (mutation) in the exact amino acid sequence in the prion protein molecule resulting in its folding configuration. Certain amino acids tend to make more efficient hinges or joints than others. It is this variability that controls which of the transmissible spongiform encephalopathies (see Table 21.1) results. For example, in Gerstmann–Straussler–Scheinker disease the amino acid valine occurs at position #117 in the prion protein molecule rather than the usual alanine.

There is as yet no cure for any of the prion diseases; however, there have been preliminary clinical studies suggesting that quinicrine may be helpful in some cases of Creutzfield–Jakob disease (see Chapter 21). This evidence is at least compelling enough that a larger collaborative study is being planned in Great Britain under the auspices of the Medical Research Council.

Wendell Stanley's crystallization of tobacco mosaic virus (Chapter 11) has challenged the conventional definition of life. The elucidation of the nature of the prion has added to the complexity of this distinction by questioning the essential features of the viruses. The prion is virus-like in being both filterable and self-reproducing. In contrast to viruses, however, it lacks nucleic acid, and so does not code for its own structure. Ultimately this ambiguity may require either a change in the classification of the prions or a change in the definition of viruses.

Part E

PARASITES

Parasitism is prevalent in the biological world because it is less energy-costly to live within the intestine or tissues of a host where the medium is the meal than to engage in, for example, predation, which requires running down and killing a prey in order to obtain food. Parasitism thus may be considered an adaptive advance over the free-living state.

The parasites that infest man range from single-celled microscopic protozoa that reproduce by simple asexual binary division to complex multi-cellular macroscopic worms in which separate male and female forms exist (Table E.1). Their life cycles can be simple, ranging from host to environment to another host, or complex, involving host to one of more intermediate forms in several different species before infecting the primary host again. The *definitive host* is defined as the one in which sexual reproduction occurs, while the *intermediate host* supports asexual reproduction.

Table E.1. The phyla of parasites that affect man, with examples of important and common disease processes.

Phylum	Class	Examples of disease
Single cell protozoa	Flagellates	Giardia diarrhea
	Pseudopods	Amebic dysentery
	Non-motile	Malaria
	Ciliates	Diarrhea
Multi-cell flatworm	Flukes	Liver flukes, blood flukes (schistosomes)
	Tape worms	Fish tape worms, pork tape worms
Multi-cell round worm	Worms	Ascaris, hookworm

The clinical diseases caused by parasites cover the entire spectrum of those caused by all other infectious agents. In the majority of cases it is not obvious whether an illness is caused by a parasite or by some other class of agents such as bacteria or viruses. Clinical disease is, however, an undesirable and collateral consequence of the basic host–parasite relationship. The ideal parasite causes little or no harm to the host, because if the latter is killed a new one must be found.

In contrast to viral, bacterial or fungal pathogens, the diagnoses in many parasite infestations relies primarily on direct visualization of the organism (either microscopically or macroscopically) rather than on laboratory culture procedures.

Control of most parasite species is an epidemiological and environmental activity directed against towards interfering with their life cycles.

Again in contrast to other classes of infections, there are only a relatively few therapeutic agents available for treatment. Further, some of these agents show significant toxicity. Development of anti-parasitic drugs is not as advanced as that of drugs for other classes of infectious agents, possibly because those areas of the world that have the greatest need for these drugs have the least financial and pharmaceutical resources for their development and deployment.

In addition to the invasive parasites of man there are ectoparasites, which live on the skin or bite but do not invade the body (Table E.2). The two major classes of ectoparasites are the spider-like and insect-like forms. The former are characterized by a body composed of two parts – the front is the cephalothorax and the rear is the abdomen – and there are four pairs of legs. The latter has a three-part body – head, thorax and abdomen – and there are three pairs of legs. Of particular importance are the ticks, mites and sucking lice that carry rickettsial infections, fleas that carry plague, and mosquitoes that transmit malaria and yellow fever. In the ticks and mites, the three body parts are fused.

Parasitic infestations are often considered to be problems only of exotic and underdeveloped regions of the world, such as parts

Table E.2.   Common types of ectoparasites of man, with examples of
diseases caused.

Phylum	Class	Types	Diseases
Arthropods	Spider-like	Scorpions	Venomous
		Spiders	Venomous
		Ticks	Rickettsia
		Mites	Rickettsia
	Insects	Sucking lice	Classical typhus
		Cone-nose bug	S. A. trypanosomiasis
		Bees, wasps, hornets, ants	Venomous
		Fleas	Bites, plague
		Mosquitoes	Malaria, yellow fever

of Africa, South America and the Orient. In fact these infestations
occur in developed areas, including the United States. Examples
include: trichomoniasis, a protozoan infestation of the genital tract
which is benign but also very common and annoying; hookworm
infestation, with its debilitating anemia, which occurs in the southern US, mostly among children who go barefoot; toxoplasmosis,
which is caused by a protozoan that can complicate pregnancy;
and pinworm infestation in children, which occurs particularly
in poor socioeconomic conditions that exist in some parts of
developed countries and is implicated in behavioral problems.

Only approximate census data are available to estimate the quan-
titative human burden of parasites. However, it would probably
not be an egregious exaggeration to estimate that on average
every person in the world today is infected by at least one parasitic
species. Certainly the world food supply could be better allo-
cated if it were used only by man and not also by his parasites.

23

MALARIA

I was tired, and what was the use? I must have examined the stomachs of a thousand mosquitoes by this time. But the Angel of Fate fortunately laid his hand on my head.

(SIR RONALD ROSS, 1857–1932)

RONALD ROSS AND CHARLES LAVERAN

Two Nobel Prizes for Physiology or Medicine were given for research on malaria; to Ronald Ross in 1902 for his elucidation of the parasite life cycle in the mosquito, and to Charles Laveran in 1907 for his discovery of the parasite in man. The awarding of two prizes for this single entity indicates the major importance of this infectious disease a century ago – an importance which, if anything, has actually increased in the world today.

Charles-Louis-Alphonse Laveran (18 June 1845–18 May 1922) was a French physician, pathologist and parasitologist. He received his medical degree in Strasbourg, and then entered the army as a surgeon practicing and teaching military medicine. In 1897 he joined the Pasteur Institute, Paris, and in 1907 he established their Laboratory of Tropical Diseases. He was a major force in advancing tropical medicine as a distinct discipline.

Ronald Ross (13 May 1857–16 September 1932) was a British physician and bacteriologist, born in India. He received his medical degree in 1879, and after serving as an Army physician he studied bacteriology from 1888 to 1889 in London. In 1892 he returned to India to begin a series of investigations on malaria with Patrick Manson. He joined the Liverpool School of Tropical Medicine and then became director of the Ross Institute and Hospital for Tropical Diseases in London, which was founded in his honor. He was knighted in 1912.

MALARIA

The term 'malaria' derives from the Italian meaning 'bad air', and refers to the malodorous swampy regions where the infecting mosquitoes abound.

Malaria is the most common severe infectious disease in the world today, with 300–500 million cases and 2–3 million deaths annually. However, tuberculosis remains the most common cause of infectious disease deaths (see Chapter 8).

In 1880, Laveran, while working in Algeria, described the malaria parasites in the blood of infected patients.

In 1899, Ross, using avian malaria as a model system, described the life cycle of the parasites in the anopheline mosquito. In particular he noted the infecting sporozoites in the salivary glands of the insect – the key feature in malaria transmission.

Malaria is transmitted man-to-man by mosquitoes (Figure 23.1). An infected female mosquito, in the process of taking a blood meal, contaminates the bite site with saliva containing malaria sporozoites. These travel through the host's bloodstream to the liver, where they mature into tissue schizonts. These liver cell schizonts divide into merozoites, which upon release invade red blood cells. In the red blood cells the merozoites are transformed into schizonts, mature into merozoites, lyse the red blood cells, and are released into the circulation (the erythrocytic cycle). This erythrocytic merozoite–schizont–merozoite cycle repeats

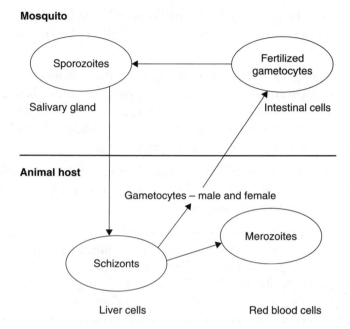

FIGURE 23-1
The life cycle of the malaria parasite.

every two days for the three tertian malarias, and every third day for the quartan malaria. Two- and three-day cycles are termed tertian and quartan respectively because the fever day is designated day one rather than the more correct day zero. It is the lysis of infected red blood cells on day 'one' that causes the fever, chills, and anemia characteristic of the infection.

Not all merozoites reinfect erythrocytes to form schizonts; some develop into gametocytes, male or female. These gametocytes are taken into the gastrointestinal tract of biting mosquitoes, where they fertilize and enter the lining cells of the insect gut. The fertilized gametocytes in the mosquito mature into motile sporozoites, migrate to the salivary glands, and are injected into the animal host with the next biting.

In the case of *P. vivax* and *P. ovale* malarias some of the schizonts remain dormant in the liver and months later mature, causing relapses of the infection. In *P. falciparum* and *P. malariae* infection there are no dormant liver schizonts, and therefore no relapses occur.

The schizonts of *P. falciparum* are sparse in the peripheral blood, as these forms tend to sequester in the peripheral vessels of the brain, kidney and placenta. The resulting cerebral malaria, bloody urine of 'Blackwater fever' and abortions all indicate the unique severity and lethality of the *P. falciparum* malaria.

An attack of malaria does not confer immunity to re-infection; however, a partial immunity does develop so that repeat infections cause less severe disease.

Malaria presents an instructive example of how an infectious disease can influence the distribution of genes in a population. An abnormal gene for hemoglobin exists in up to 40 percent of Africans. This abnormal red blood cell hemoglobin does not transport oxygen well. Also these cells do not survive normally, which results in anemia. The cells also take on an elongated shape and hence are termed 'sickle cells'. Sickle cell hemoglobin persists in the population, despite its defects, because the red blood cells do not support the growth of malaria parasites well. For this reason persons with sickle cell hemoglobin have increased resistance to malaria and therefore some selective advantage over the population with the normal hemoglobin. It could be speculated that if malaria were ever eradicated, then over the course of many generations sickle cell anemia might disappear entirely.

Control and treatment

The control of malaria is based upon suppression of mosquito populations and upon prophylactic administration of antimalarial drugs. Because mosquitoes bite most frequently in the evening, mosquito control measures include remaining indoors in the evenings and using protective netting at night. Currently bed netting coated with permethrin, an insect repellent, is being tested in Kenya. In addition, pools of water where the mosquito larvae develop are drained. Anti-mosquito sprays are also effective.

The most effective spray, DDT, which was developed for this specific purpose, is no longer used because of ecologic considerations (see Chapter 25).

The first effective drug for treatment of malaria was quinine, derived from the bark of the cinchona tree. Quinine was discovered by an Argentinian monk in Peru, who in 1633 described the use of the powdered bark of the cinchona tree for the treatment of 'fevers and tertians'. The cinchona tree, a native of the South American Andes, was exported and widely cultivated in Java, India and Sri Lanka for quinine production.

Folklore has it that the ability of quinine to prevent and treat malaria, as well as other non-malarial fevers, accounted for the popularity of quinine water among the colonists in the tropical regions of the world.

Quinine was the primary antimalarial drug up to the Second World War, when it was replaced by chloroquine – the first synthetic antimalarial. With widespread use, however, chloroquine-resistant strains of malaria appeared, resulting in the need to develop additional drugs.

The several stages of the malaria parasites in man (see Table 23.1) have differing susceptibilities to antimalarial drugs. For example, to treat a clinical case of malaria it is necessary to use an agent that inhibits the parasites in the red blood cells (erythrocytic cycle) because it is the infection and lysis of these cells that causes clinical manifestations of the disease. Both quinine and chloroquine are useful for this purpose. To prevent a later relapse, however, it is necessary to kill the latent parasites that remain in the liver (the exerythrocytic cycle), a component of *P. vivax* and *P. ovale* infections. Neither quinine nor chloroquine is effective against the exerythrocytic forms, and so an agent such as primaquine must be added to prevent relapses. This principle applies both when antimalaria drugs are given to prevent as well as when they are given to treat the infection. Visitors to overseas countries are familiar with the two-pill regimen for malaria prophylaxis; chloroquine to prevent the erythrocyte cycle and primaquine to eliminate the exerythrocytic cycle.

Table 23.1.   Malaria parasites of man.

Plasmodium species	Common name	Fever cycle (days)	Geographic distribution	Relapses	Lethality
Vivax	Benign tertian	2	Central America, South America, India subcontinent, Central Asia	Frequent	Mild
Falciparium	Malignant tertian	2		None	Severe
Ovale	Ovale	2	Subtropics, Sub-Saharan Africa, Papua New Guinea	Moderate	Mild
Malariae	Quartan	3	Widely scattered	None	Moderate

This principle was also applied in testing drugs for antimalarial efficacy in man during the Second World War, when there was a risk that supplies of quinine could be curtailed. Volunteers were inoculated with blood from malaria patients which, containing only merozoites, caused an infection restricted to the red blood cells. As the inoculum did not contain sporozoites, which occur only in mosquitoes, no exerythrocytic cycle was initiated and consequently there was no risk of relapse after the acute infection was treated.

Malarial parasites can develop resistance to antimalarial agents, as has occurred with chloroquine, which is no longer useful for treatment of falciparum infections. Resistance has necessitated the development of new synthetic agents, some of which are based upon the principle pioneered by Elion and Hitchings (see Chapter 7).

An interesting new class of agents, which have been under investigation in clinics for the past two decades, has been isolated from the sweet wormwood, *Artemesia annua*, a plant that has been used by Chinese practitioners for centuries to treat fevers. This class is particularly effective in the treatment of *P. falciparium* cerebral malaria.

The sources of antimalaria drugs seems now to have come full circle, starting with quinine from the cinchona tree, progressing through a number of laboratory-synthesized chemotherapeutic agents and now artemisias extracted from sweet wormwood. A vote of confidence for alternative medicine, or simply a pragmatic approach to therapeutics?

A number of laboratories are currently attempting to develop a malaria vaccine, but to date the goal has proved elusive. If successful, however, an effective and practicable malaria vaccine could merit an unprecedented third Nobel Prize for research in the same infectious disease.

24

CANCER PARASITE

The medical errors of one century constitute the popular faith of the next.

(ALONZO CLARK, 1907–1887)

JOHANNES FIBIGER

Johannes Fibiger received the 1926 Nobel Prize for Physiology or Medicine for the first controlled laboratory production of cancer in experimental animals.

Johannes Andreas Grit Fibiger (23 April 1867–30 January 1928) was a Danish pathologist. He received his MD degree in 1890, and then traveled to Berlin where he studied bacteriology with Emil von Behring (see Chapter 1) and Robert Koch (see Chapter 8). He returned to Denmark in 1894 to join the faculty of the University of Copenhagen, and in 1900 became Professor of Pathological Anatomy and Head of the Institute of Pathologic Anatomy.

FIBIGER'S RESEARCH

Fibiger's studies on cancer began with his observation of a gastric growth, apparently malignant, in three laboratory rats that had been infected with the tubercle bacillus. In the stomach of these rats he found a parasitic worm *Spiroptera neoplastica* (*Gongylonema neoplastica*). To further investigate this phenomenon he then fed rats these adult parasitic worms plus their eggs, but was unable to reproduce the gastric lesions. In the normal life cycle of these worms the eggs are eaten by cockroaches; they then hatch, producing larvae that encyst in the muscles of the insect. Upon feeding rats larvae-infected cockroaches, the previously seen growths were reproduced.

This initial work appeared to be an entrée into the controlled laboratory studies on the cause and course of cancer. However, further studies of the parasite-infected rats produced no evidence of any metastatic cancer in their other organs, and additional histologic studies of the gastric growths raised doubts as to whether they were actually cancers.

Fibiger's Nobel Prize is often cited as one that was probably given prematurely, and may have been the reason why no Nobel Prizes for cancer research were given until forty years later. While these misgivings may have been justified, with the clearer perspective of hindsight and additional research it now appears that Fiberger's basic concept had some merit.

There are well-documented examples of cancers in man associated with parasitic infestations. *Schistosoma haematobium*, a blood fluke which occurs in Africa and the Middle East, infests the small veins of the urinary bladder and produce eggs that are passed in the urine. These eggs develop in water-dwelling snails, from which larvae are released that penetrate the skin of the human host and migrate to the bladder venules. In addition to causing blood loss in the urine and scarring of the bladder, the infestation doubles the risk of bladder cancer. In Egypt, this accounts for 16 per cent of all such cases.

In addition to cancer associated with blood flukes, neoplasia is also associated with the liver flukes *Opisthorcis sinensis* and *O. viverrini*, which occur in the Far East. The adult flukes inhabit the bile ducts of the liver. These eggs hatch in water and infect fish, which are eaten by man and develop into adult flukes, which take up residence in the liver, thus completing the cycle. The liver infestation not only results in scarring and obstruction of the bile ducts, but is also associated with an increased risk of bile duct cancer, *cholangiocarcinoma*.

THE RELATION BETWEEN INFECTION AND CANCER

A line of research carried out during the past two decades has added an important new dimension to the relation between infection and cancer. In 1984, two Australians workers, Marshall and Warren, reported that a bacterium, *Helicobacter pylori*, caused gastric inflammatory disease and ulcers. The organism occurs in the stomach of 90 per cent of individuals in developing countries and 50 per cent of those in developed countries of the world. *H. pylori* resides within the mucus that coats the inner wall of the stomach. The bacteria can persist despite the hostile high acidity environment because of their production of the enzyme urease, which converts urea to ammonia – a highly basic chemical which locally neutralizes the gastric acidity.

H. pylori has now been associated with four important diseases of the stomach: gastritis, peptic ulcer, MALT lymphoma and adenocarcinoma.

Gastritis is a painful inflammation of the stomach lining and wall which can be acute or chronic and can eventually lead to achlorhydria, a failure of the normal acid production in the stomach which is necessary for digestion.

Peptic ulcers are local areas of erosion that bleed, causing anemia, and can perforate completely through the wall of the stomach into the abdominal cavity, precipitating a surgical emergency. These ulcers are in the stomach and also in the duodenum, the initial segment of the small intestine that attaches to the stomach and receives the partially digested food. The two most common causes of gastritis and peptic ulcer are *H. pylori* infection, and the use of aspirin and similar drugs for pain and arthritis. The former accounts for about three-quarters of all cases, and the latter for the remaining one-quarter.

Of particular relevance to the early work of Fibiger are the two neoplastic disorders that are associated with *H. pylori* infection: MALT lymphoma and adenocarcinoma. MALT tumors of the stomach are a form of lymphoma occurring in the gastric *mucosal-associated lymphoid tissue*. *H. pylori* MALT tumors account for about 5 per cent of all gastric malignancies. These tumors are unique in that they appear to develop as a result of antigenic stimulation by *H. pylori*, so they will resolve if the infection is treated with antibiotics early in the infection but not later, after they become autonomous.

Gastric adenocarcinoma is also statistically associated with *H. pylori* infection. It has long been recognized that gastric achlorhydria, which can be caused by gastritis, is a risk factor for gastric adenocarcinoma, and this association may play a role in the development of *H. pylori* malignancy. In addition there is epidemologic evidence that the current decreasing incidence of *H. pylori* infection due to antibiotic treatment has led to a decreasing incidence of this gastric malignancy.

Twenty years ago peptic ulcer disease, which occurs in 1–2 per cent of the general population, was held without question to be a metabolic disorder related to excess stomach acid and the stomach enzyme *pepsin*. The studies with *H. pylori*, however, have revolutionized the understanding of the cause and therapy of this disorder. Further, an association between infection and gastric malignancy was never seriously entertained before the *H. pylori* studies, and the cure of some forms of gastric malignancy with antibiotics was without precedent. Because of this revolution in gastroenterology resulting from the work with *H. pylori*, its discovery must certainly be on everyone's shortlist of potential Nobel Prize-winning research.

The contribution of Fibiger's work was not the discovery of the specific cause of cancer; rather it changed cancer from a study of spontaneously arising tumors in animals to a controlled induction of these lesions, greatly facilitating investigations on the cause, course and management of neoplasia in general. Further, epi-demiological data on the parasites *S. haemotobium*, *O. sinensis*, and *O. viverrini*, and on the bacterium *H. pylori*, have estab-lished a relationship between gastric malignancy and infectious agents, although not Fibiger's *G. neoplastica*.

Fibiger may have not got it all right, but he certainly did not get it all wrong.

25

DDT

DDT went further toward the eradication of malariologists than of mosquitoes.

(RENÉ J. DUBOS, 1901–1982)

PAUL MÜLLER

Paul Müller received the 1948 Nobel Prize for Physiology or Medicine for his discovery of the insecticidal properties of dichlorodiphenyltrichloroethane (DDT).

Paul Hermann Müller (2 January 1899–12 October 1965) was a Swiss chemist at the J.R. Geigy Company, Basel, from 1925 to 1965. He sought the 'ideal' insecticide, which he defined as having four characteristics: rapid action against a wide variety of insects, non-toxicity to plants and animals, chemical stability, and economical production and application. His candidate, which he synthesized in 1939, was DDT. This compound had previously been synthesized by a German chemist, Othmar Zeidler, in 1874; however, its insecticidal property was not recognized at the time.

DDT

DDT is an organic halogen that acts by disrupting the function of the nervous system, and kills insects rapidly upon contact.

In 1939 the Swiss government tested DDT against the Colorado potato beetle, a crop blight of worldwide importance. In 1943, the US Department of Agriculture did the same. In 1944 it was used to abort an outbreak of louse-borne typhus (see Chapter 9) in Naples, Italy. Because DDT proved to be remarkably successful in these initial trials, it was quickly accepted and used worldwide. During the Second World War, DDT was initially employed as a mothproofing in clothing and then as an insecticidal dusting powder, both for military combatants and for displaced civilians. It was found to be effective against lice, the vector of typhus; fleas, the vector of plague; and mosquitoes, the vector of both malaria (Chapter 23) and yellow fever (Chapter 12). In addition to preventing infectious disease transmission, its use in agriculture markedly increased crop yields and food supplies.

PROBLEMS WITH DDT

The problem with DDT eventually proved to be not any acute toxicity to man but rather its harm to all of the other biota of our planet. Questions and concerns about the medical and environmental safety of DDT were first raised by the author Rachel Carson, the biologist and ecologist, in her 1962 sentinel volume *Silent Spring*. She pointed out that the Achilles heel of the insecticide was that it too successfully fulfilled one of Müller's four criteria – chemical stability. DDT persisted in the environment in overwhelming amounts. In 1968 it was estimated that there were half a million tons of DDT in the environment. Quantities ranging from 10 to 100 pounds per acre were found.

As DDT is not biodegradable, it tends to become progressively more concentrated as it moves up the food chain. It is taken in by insects, and then passed to insect-eating birds and fish, and on to higher predators. As DDT is a fat-soluble chemical and so tends to accumulate in the fat deposits of the body, it is particularly devastating to meat-eating birds such as falcons, hawks and eagles, and to fish-eaters such as pelicans, cormorants and egrets. DDT toxicity results in the production of eggs with soft and fragile eggshells that cannot be successfully incubated – an effect that accounts for the observed decline in bird populations world-wide, especially in North America, Europe and Japan. In addition there was also the suspicion that DDT was harmful to sea plankton, upon which all fish ultimately rely for their primary food supply.

Toxicity of chronic environmental exposure to DDT has not been established in man, although the agent has been detected in food for human consumption as well as in human breast milk. Acute DDT toxicity in man has been described, and includes nausea, irritability, weakness, muscle tremors and convulsions. There is no specific antidote for this syndrome.

Another problem associated with the widespread use of DDT, and one that should not have been unanticipated, was the development of insect resistance to the insecticide. This has been observed particularly in house flies and mosquitoes, requiring substitution of other insecticides for their eradication.

Because of the health and environmental hazards of DDT, its use was restricted by the US Environmental Protection Agency in the 1960s and it was almost completely banned in 1972 – a move that was subsequently adopted by most other governments worldwide. Recently, however, some social observers have argued that the health benefit:hazard ratio in certain developing countries warrants the reintroduction of this insecticide, both to combat insect-transmitted disease and to increase the available food supply. Even Rachel Carson in *Silent Spring* recognizes that:

> disease-carrying insects become important where human beings are crowded together, especially under conditions where sanitation is poor, as in time of natural disaster or war, in situations of extreme poverty and deprivation. Then control of some sort becomes necessary.

The pro-DDT advocates ignore the environmental harm of the insecticide, while Ms Carson might have argued that insect 'control' does not necessarily imply DDT specifically.

Work on improved insect control measures continues, including not only better insecticides but also genetically modified insect-resistant crops and novel environmentally friendly biological control measures such as stocking ponds with small fish that feed on insect larvae.

Müller's search for the 'ideal' insecticide continues. It will be difficult to find another insecticide as good or as bad as DDT.

COUNTERPOINT: HIV/AIDS

Disease has social as well as physical, chemical, and
biologic causes.

(HENRY E. SIGERIST, 1891–1957)

INTRODUCTION

The discovery of the etiology and pathogenesis of HIV/AIDS is arguably the most important advance in infectious diseases that has been made since the Nobel Prizes were first awarded over a century ago. The Nobel Committee, however, has not to date chosen to recognize this achievement, despite the fact that the basic research was done over twenty-five years ago. Further, for reasons described below, it seems possible that the study may never be so recognized.

HIV refers to the Human Immunodeficiency Virus, the cause of the infection, and AIDS to the Acquired Immunodeficiency Syndrome, its immunological consequence.

The saga of the current catastrophic worldwide pandemic of HIV/AIDS began in 1981 with a report in the CDC's *Morbidity Mortality Weekly Report* of pneumocystis pneumonia occurring in five homosexual men living in the Los Angeles area. Pneumocystis pneumonia is an infection that characteristically occurs in malnourished infants living in impoverished conditions, and in individuals with impaired immunity. It seemed likely therefore that the five men had all been subjected to some influence which had damaged their basic immune defense mechanisms. The discovery of the virus that caused this immuno-suppression and the methods for diagnosis and treatment of the infection are triumphs of modern medical research.

THE WORLDWIDE PANDEMIC

HIV/AIDS, first recognized in homosexual men, has evolved to the current worldwide pandemic which now includes heterosexuals and children. The main risk factors are now unprotected sex, recreational drug injections, and infected mother-to-offspring transmission.

Currently the scope of the pandemic is enormous and appears to be increasing, despite knowledge of its epidemiology and of effective control measures. The statistics are staggering. There are now 40 million cases of HIV/AIDS worldwide; 3 million people die of the disease and 5 million new cases are added each year. Sub-Saharan Africa is currently the epicenter, with 5 million cases and 2.3 million deaths, and 3.2 million new cases, per year. In Francistown, Botswana, 50 per cent of pregnant women and 40 per cent of all adults are HIV positive. The figure for adults in Swaziland is 39 per cent, and in Zimbabwe is 33 per cent.

Most recently the pandemic is spreading to Asia and Europe. One in four new infections is occurring in Asia. There are 4.5 million new cases in India. In China, 1–2 million cases have appeared, and this number is increasing by 30 per cent annually. Further, the fastest growing epidemic area is now developing in Eastern Europe, which had previously been relatively disease-free.

Worldwide, the expenditures by governments to control HIV/AIDS amounts to five billion dollars annually. Despite this significant expenditure it is estimated that the minimum amount needed for the effort is ten billion dollars annually.

ORIGIN OF THE HIV VIRUS

The origin of the HIV virus has been a matter of extensive conjecture, but the current consensus is that it evolved in Africa as a result of the transmission of the Simian Immunodeficiency Virus (SIV) from infected chimpanzees to man. It is unknown when this might have occurred, but it is most likely decades, even centuries, ago. HIV could well have been endemic in Africa for many years and escaped recognition as a unique clinical entity because the natural history of the infection is that of years of slowly progressive weakness, anorexia and wasting, and eventually death from an intercurrent infection such as tuberculosis. This constellation of signs and symptoms is actually that of the 'thin disease', which has been recognized in Africa for many years. Also, HIV/AIDS illness could easily in the past have been attributed to the non-specific effects of poor social and sanitary conditions, lack of adequate medical services, and chronic famine conditions. However, now that the disease has been virologically, serologically and immunologically defined and has become pandemic, it is readily recognized.

PREVENTION AND TREATMENT

Sexual transmission is prevented primarily by condom use. The risk from illicit drug use is needle-sharing among addicts, and one method of control, besides withdrawal treatment, is the provision of various needle exchange programs. Maternal-to-newborn infection can occur either transplacentally or via breastfeeding, and can be controlled by short-term antiviral treatment of the mother and newborn and by using bottle-feeding. Transmission via blood transfusion, which previously accounted for the high incidence in hemophiliacs, is no longer a significant risk, as blood and blood products are routinely screened for the HIV virus.

The discovery of the HIV viral etiology of AIDS was accomplished by two groups in the 1980s; the teams led by Dr Robert Gallo at the National Institutes of Health, Bethesda, Maryland, and by Dr Luc Montegnier at the Pasteur Institute, Paris. The agent, termed Human Immunodeficiency Virus, was found to be a member of the retrovirus group (see Table 19.1). With the discovery of the viral etiology of AIDS, antiviral therapy was soon introduced and widely applied. Currently there are two classes of antiviral agents in use; the reverse transcriptase inhibitors, which interfere with the enzyme needed for viral nucleic acid synthesis, and the protease inhibitors, which block protein synthesis. These drugs in various combinations constitute HAART (highly active antiretroviral therapy). Although there is as yet no cure for HIV/AIDS, HAART has markedly increased the quality of life and longevity of these patients. In addition, research on the development of an HIV vaccine is being pursued actively in both academic and pharmaceutical centers worldwide.

A NOBEL PRIZE?

Although HIV/AIDS is of monumental international medical and social importance, there has been no Nobel Prize awarded for the work on it. This may be the result of the initial controversy between the two groups who unraveled the etiology of this pandemic. There was also some uncertainty regarding the origin of a strain of HIV that was studied in both laboratories. Currently, however, a rapprochement seems to have been reached between the two groups, as indicated by two summary articles coauthored by Gallo and Montegnier, one appearing in the *Scientific American* in October 1988 and the other in the *New England Journal of Medicine* on December 11, 2003. These papers indicated that the first isolate of an HIV was made by the French group in 1983, and the first production of the virus in sufficient quantities to examine the biology of the virus and develop diagnostic serologic tests was performed by the US group. Further, the authors indicated that a strain of HIV that had been of uncertain history had actually been first isolated in the Paris laboratory but occurred admixed with HIV strains in both laboratories due to technical problems. The 1988 joint article stated:

> Thus contributions from our laboratories – in roughly equal proportions – had demonstrated that the cause of AIDS is a new human retrovirus.

The above uncertainty, although apparently laid to rest by the two primary workers, may be the reason for the Nobel Committee's reluctance so far to enter the picture. The ultimate disposition of the matter by the Nobel Committee is a matter for the future to disclose; however, it should be noted that it is now a quarter-century since the original reports.

It should be recognized that the selection of Nobel Laureates is a human endeavor, and for this reason the occasional introduction of some social considerations into the objective scientific process might not be totally unexpected.

It is ever thus.

INDEX

Acquired immune system, 11–12, 13
Acquired Immunodeficiency Syndrome, 155
  see also HIV/AIDS
Actinomycin, 48
Acyclovir, 35, 52, 54
Aerobic bacteria, 56
Allergic reactions, 11, 25
Alpers, 124
Alpharetroviruses, 116, 117
Amebic dysentery, 59
Aminoglycosides, 49
Anaerobic bacteria, 19, 56
Aniline dyes, 61
Anthrax, 4, 59
Antibiotics, 2, 35, 48
  from soil microorganisms, 48–50
  new agent identification methods, 52
  resistance, 4, 48–9
Antibodies, 12–13, 24–6
  clonal theory, 26
  production, 26, 33
  specificity, 12–13
  see also Immunoglobulins
Antibody-mediated immunity, 13
Antigens, 11
  antibody production stimulus, 26
  antibody specificity, 12–13
  MHC/HLA system, 31–2, 33
  presentation to immune system, 33, 34
Antimalarial drugs, 140–2
  new agents, 142–3
  prophylaxis, 142
  resistance, 141, 142
Antimicrobial agents, 35–54
  antagonism, 36
  cidal, 36
  mechanisms of action, 35
  resistance, 36–7
  static, 36
  synergism, 36
  therapeutic index, 35, 36
  see also Chemotherapeutic agents
Antimicrobial defenses, 23–8
Anti-parasitic agents, 134
Antiserum (antitoxin serum), diphtheria, 16
Antiviral agents, 54, 79
  hepatitis B, 97, 98
  hepatitis C, 99
  HIV/AIDS, 157–8
  reverse transcriptase targeting, 117

Arsphenamine, 44, 75
Artemesia annua, 143
Aspirin, 73, 147
Atypical mycobacteria, 66
Australia antigen, 96
Azo dyes, 40

B-cells, 12, 13, 25–6
  antibody production, 26, 33
  T4 helper cell actions, 33
Bacilli, 55
Bacteria, 35, 36, 55–75
  antimicrobials resistance, 36–7
  biological characteristics, 57
  classification, 55
  dihydrofolate reductase, 53
  diseases, 57
  exo/endotoxins production, 11, 57
  Gram staining, 55–6
  heat killing, 56
  laboratory culture, 56
  lysogenic conversion, 108
  multiplication, 56–7
Bacterial conjunctivitis, 59
Bacteriophage, 77, 80, 103–5
  host cell entry, 78
  lysogeny, 107–8
  virulent infection, 108
Ballistite, 6
Baltimore, David, 115–16
Barley yellow dwarf virus, 83
BCG vaccine, 65
Beijerinck, Willem, 82
Bile duct cancer (cholangiocarcinoma), 147
Bioterrorism/biological warfare, 4, 88, 93
Bird flu, 3
Bishop, Michael, 119
BK virus, 113
Black, James, 51, 142
'Blackwater fever', 140
Bladder cancer, 146
Blood, 23–4
Blumberg, Baruch, 95, 96
Bone marrow transplantation, 13
Bordet, Jules, 23, 24, 28
Bordetella pertussis, 28
Botulism, 56
Bovet, Daniel, 40
Bovine spongiform encephalopathy (mad cow disease), 3, 125–6, 128

Brucellosis, 67
Burnet, Macfarlane, 26

Calmette, Albert, 65
Cancer
  infection relationship, 147–9
  parasite infestation relationship, 145–6
Caroll, James, 87
Carson, Rachel, 152, 153
Cell-mediated immunity, 13
Chain, Ernest, 43–4, 45, 46
Chase, Martha, 104
Chemotherapeutic agents, 35, 51–4
  new agent identification methods, 52
  see also Antimicrobial agents
Chloroquine, 141, 142
Cholangiocarcinoma (bile duct cancer), 147
Cholera, 59
Chronic wasting disease of elk and deer,
  126
Cinchona tree, 141, 143
Cirrhosis, 97, 98, 99
Citrus quick decline virus, 83
Clones, 56, 61
Clostridia, 56
Clostridium tetani, 19
  lysogenic conversion, 108
Cocci, 55
Cold sores, 54
Complement, 11, 26–8
  cascade, 27
Cordite, 6
Corneal ulceration, 54
Corona virus, 3
Corynebacterium diphtheriae, 16
  lysogenic conversion, 108
Cowpox (vaccinia), 1
Creutzfeldt, Hans Gerhard, 127
Creutzfeldt–Jakob disease (CJD), 126,
  127–8, 131
  new variant (vCJD), 128
Cytomegalovirus, 78
Cytotoxic T-cells see T8 suppressor/cytotoxic
  cells

Darwin, Charles, 34
DDT, 68, 141, 151–4
  environmental persistence, 152–3
  health benefit:hazard ratio, 153
  insect resistance, 153
  insecticidal activity, 151–2
  toxicity, 152–3
Deltaretroviruses, 116, 117
d'Herelle, Fi, 104
Dihydrofolate reductase, 52, 53
Diphtheria, 15–19, 21, 67
  active immunization, 16, 20
  antiserum (antitoxin serum), 16
  1925 Nome (Alaska) serum run, 17–19
  toxin, 16, 57, 108
DNA, 105

DNA polymerase, 54
DNA tumor viruses, 111, 112–13, 116
Doherty, Peter, 29–30, 33
Domagk, Gerhard, 39–40
Dreiber, 92
Dulbecco, Renato, 111, 112, 115, 116
Dynamite, 6
Dysentery, 57
  amebic, 59

Ebola virus, 3
Ectoparasites, 134, 135
Ehrlich, Paul, 26, 61, 75
Elion, Gertrude, 51, 52, 53, 54, 142
Encephalitis, 30, 54
Enders, John, 89–90, 94
Endocytosis, 78
Endotoxins, 57
Ericsson, John, 5
Erythrocytes (red blood cells), 23
Escherichia coli T bacteriophages, 104
Eukaryotes, 55
Exotoxins, 57

Fatal familial insomnia, 126
Fever, 73
Fever cabinets, 73
Fibiger, Johannes, 145, 146, 148, 149
Fildes, P., 41
Finley, Carlos, 86
Flatworms, 133
Fleas, 134
  DDT in control, 152
  endemic typhus transmission, 70
Fleming, Alexander, 43, 44
Florey, Howard, 43, 44, 45
Flukes, 146–7
Folic acid metabolism, 41, 52–3
Food poisoning, 57
Fowl sarcoma, 110
Fungal infection, 35

Gajdusek, Carleton, 123, 128
Gallo, Robert, 157, 158
Gamma globulins, 24
Gas gangrene, 56, 57
Gastric adenocarcinoma, 147, 148
Gastritis, 147
General paresis, 75
Genetic information, 105
Genetically modified food crops, 108
Gengou, Octave, 23, 28
Gerstmann–Strausseler–Scheinker
  syndrome, 126, 131
Gibbs, 124, 128
Global Fund, 2
Gongylonema neoplastica, 146, 149
Gorgas, W.C., 87
Gould, Stephen Jay, 57
Gram, Hans Christian Joachim, 56

Gram-negative bacteria, 56
Gram-positive bacteria, 56
Granulocytes, 24
Guerin, Camille, 65

HAART (highly active antiretroviral
   therapy), 158
Hansen's bacillus (leprosy bacillus), 66
Helicobacter pylori, 147–8, 149
Helminth defenses, 25
Helper T-cells *see* T4 helper/inducer cells
Hemophilus meningitis, 40
Hepatitis, 95–6
   viral characteristics, 100
Hepatitis A, 96, 98–9, 100
Hepatitis B, 79, 95–101
   Australia antigen, 96
   chronic liver disease, 97, 98
   specific immune serum, 97, 98
   transmission, 97
   vaccination, 97–8
Hepatitis C, 79, 99–100
Hepatitis D, 100–1
Hepatitis E, 100, 101
Hepatitis non-A non B, 99
Hepatocellular carcinoma, 97, 98, 99, 100
Herpes simplex virus type 1, 54, 78
Herpes simplex virus type 2, 78
Herpesviruses, 54, 78
Hershey, Alfred, 103, 104
Hershey–Chase experiment, 104–5
Hinshelwood, Cyril, 49
Hitchings, George, 51, 52, 53, 54, 142
HIV/AIDS, 2, 3, 155–9
   antiviral agents, 117, 157–8
   associated syphilis, 75
   associated tuberculosis, 60, 157
   origins of virus, 156–7
   pandemic, 155, 156
   studies of viral etiology, 157, 158–9
   transmission prevention, 157
Hoffman, Eric, 74
Homologous serum jaundice *see*
   Hepatitis B
Hookworm, 135
House flies, 153
Human Immunodeficiency Virus, 117, 155
   origins, 156–7
   see also HIV/AIDS
Human Leukocyte Antigen (HLA), 31
   alleles, 32, 33
   Class I antigens, 32
   Class II antigens, 32–3
Human louse (*Pediculus hominis*), 69
Hyperpyrexia, 73

Immune recall, 12
Immune specificity, 12
Immune system, 11–13
   acquired, 11–12, 13
   innate, 11

Immunity, 11–34
   antibody-mediated, 13
   cell-mediated, 13
Immunoglobulin A (IgA), 25
Immunoglobulin D (IgD), 25–6
Immunoglobulin E (IgE), 25
Immunoglobulin G (IgG), 24
Immunoglobulin M (IgM), 24
Immunoglobulins, 24–6
   classes, 25
   see also Antibodies
Infectious hepatitis *see* Hepatitis A
Influenza, 79
   bird flu, 3
   vaccine, 81
Innate immune system, 11
Interferon, 79, 99, 100
Interleukins, 33
Isaacs, 79
Ivanovsky, Dmitry, 81

Jakob, Alfons Maria, 127
JC virus, 113
Jenner, 1

Kaposi's sarcoma-associated virus, 78
Kassen, Gunnar, 17, 18, 19
Kenny, Elizabeth, 91
Keratitis, 54
Koch, Robert, 59, 60, 61, 62, 145
Koch's Postulates, 61
Kuru, 123–4
Kuru-like diseases, 124–8
   in animals, 124–6
   in man, 126–8

Lamivudine, 97, 98
Laveran, Charles, 137, 138
Lazear, Jesse, 87
Lederberg, Joshua, 56
Legionella, 3
Legionnaires' disease, 3
Lentiviruses, 117
Leprosy, 66
Leprosy bacillus (Hansen's bacillus), 66
Leukemias, 110, 116
Lice, 134
   DDT in control, 152
   human (*Pediculus hominis*), 69
   rickettsiae transmission, 69
   typhus transmission, 67, 68–9, 152
Lima, Rocha, 68
Lindemann, 79
Lockjaw *see* Tetanus
Loeffler, Frederick, 16
Lübeck disaster, 65
Lwoff, André, 107, 108
Lymphocytes, 12
Lymphocytic choriomeningitis, 29, 30
Lymphomas, 110
   MALT, 147, 148

Mad cow disease *see* Bovine spongiform
    encephalopathy
Major histocompatability complex, 30–3
    Class I genes, 32
    Class II genes, 32
    restriction, 29–34
Malaria, 2, 35, 53, 59, 134, 137–43, 152
    antimalarial drugs, 140–2
    'blackwater fever', 140
    cerebral, 140, 143
    clinical features, 139
    life cycle stages in man, 138, 140
    life cycle stages in mosquito, 137, 138–40
    mosquito control, 140–141
    neurosyphilis treatment (malariotherapy),
        71, 72
    prophylaxis, 142
    quartan, 139
    relapse, 140, 141
    sickle cell hemoglobin-related resistance,
        140
    tertian, 139
    vaccines, 143
MALT lymphoma, 147, 148
Mammary tumors, 110
Marburg virus, 3
Marshall, 147
Mayer, Adolph, 82
Measles, 67, 89
Medawar, Peter, 31
Membrane attack complex (MAC), 27, 28
MHC *see* Major histocompatability
    complex
Miller, S.F., 84
Mites, 134
Montegnier, Luc, 157, 158
Mosquitoes, 134
    control measures, 87–8, 140–1
    DDT in control, 152
    DDT resistance, 153
    malaria parasite life cycle, 137, 138
    yellow fever transmission, 85, 86–7
Mucous membranes, 25
Müller, Paul, 151, 154
Multiple sclerosis, 79
Mumps, 94
Mutation, 49
Mycobacterium tuberculosis, 48, 59, 61
    'acid-fast' staining, 61–2
    BCG vaccine, 65
    multi-drug resistance, 60, 64, 65
    streptomycin sensitivity, 64

Neomycin, 49–50
Neosalvarsan, 72
Neurons, 30
Neurosyphilis, 74, 75
    malariotherapy, 71, 72
Nicolle, Charles, 67, 68–9, 70
Nitroglycerine, 5–6

Nits, 69
Nobel, Alfred Bernhard, 5–6
Nobel Committees, 7
Nobel Foundation, 6, 7
Nobel prizes, 6–9
    laureate selection process, 8
    Nobel's will, 6–7
    subjects, 7, 8
Norwegian Nobel Committee, 7
Nucleic acids, 77, 78, 105

Oncogenes, 110, 113, 119–21
Opisthorcis sinensis, 146, 149
Opisthorcis viverrini, 146, 149
Opsonization, 27
Organic compounds synthesis, 84

Papilloma viruses, 112, 113
Papova virus, 112
Para-amino salicylic acid, 64
Parasites, 35, 133–54
    definitive host, 133
    intermediate host, 133
    life cycles, 133, 134
Passive immunization
    diphtheria, 16
    hepatitis B, 97, 98
Pasteur, 1, 2
Pediculus hominis (human louse), 69
Penicillin, 2, 35, 36, 43–6, 48, 75
    allergic reactions, 45
    mechanism of action, 45
    resistance, 46
Penicillinase, 46
Penicillium notatum, 44
Peptic ulcer, 147, 148
Permethrin, 141
Pertussis (whooping cough), 28
Petri dishes, 61
Petri, Julius Richard, 61, 82
Phagocytic cells, 11
Phagocytosis, 24, 26, 32
Pinworm, 135
Plague, 59, 134, 152
Plant-infecting viruses, 83
Plasma, 24
Plasmodium falciparum, 72, 140, 143
Plasmodium malariae, 140
Plasmodium ovale, 140, 141
Plasmodium vivax, 72, 140, 141
Platelets, 24
Pneumocystis pneumonia, 155
Pneumonia, 89
Poliomyelitis, 80, 88, 89–94, 116
    eradication program, 93
    in vitro tissue culture of virus, 89, 90, 94
    management, 91
    spread, 90–91
    symptoms, 90–91

vaccination, 92–3
viral serotypes, 92, 93
Polyoma virus, 111–13, 116
Primaquine, 141, 142
Prions, 84, 129–31
    abnormal folding configuration, 130, 131
    normal protein, 130
    nucleation-growth, 131
Progressive multifocal encephalopathy, 113
Prokaryotes, 55
Prontosil, 39–40
Protease inhibitors, 158
Proto-oncogenes, 119
Protozoa, 133, 135
Prusiner, Stanley, 129, 130
Pyrimethamine, 52–3
    mechanism of action, 52–3
    sulfonamide combinations, 53

Quinacrine, 131
Quinine, 72, 141, 142, 143

Rabies vaccination, 1, 2
Red blood cells (erythrocytes), 23
Reed, Walter, 86
Respiratory infections, 53
Retroviruses, 116, 157, 158
Reverse transcriptase, 98, 115–17
Reverse transcriptase inhibitors, 158
Rheumatic fever, 57
Ribavirin, 99, 100
Ricketts, Henry Taylor, 69, 70
Rickettsia prowazekii, 68, 70
Rickettsia typhi, 70
Rickettsiae, 68
    infections, 68, 69, 134
    transmission by lice, 69
RNA tumor viruses, 110, 113, 116, 120
Robbins, Frederick, 89–90, 94
Rocky Mountain Spotted Fever, 69
Ross, Ronald, 137, 138
Roundworms, 133
Rous, Peyton, 109–10
Rous sarcoma virus, 109–10, 115, 116
Rubella (German measles), 90, 94
Rubeola (red measles), 94

Sabin oral polio virus vaccine (OPV), 92, 93
Salk inactivated polio vaccine (IPV), 92–3, 113
Salvarsan, 75
Sarcomas, 116
SARS (Severe Acute Respiratory Syndrome), 3
Schaudin, Fritz, 74
Schistosoma haematobium, 146, 149
Scrapie, 124–5, 129–30
    abnormal prion protein, 130–1
Seppala, Leonhard, 17, 18
Serotherpay, 15–21

Serum, 24
Severe Acute Respiratory Syndrome (SARS), 3
Severe Combined Immune Deficiency (SCID), 13
Shannon, Bill, 17
Shaw, 92
Sickle cell hemoglobin, 140
Sigurdsson, B., 124
Simian Immunodeficiency Virus (SIV), 157
Simian virus 40 (SVU40u), 112–13
Sleeping sickness, 59
Smallpox, 4, 88
    vaccination, 1, 4, 88
Snell, George David, 31
Sobrero, Ascanio, 5
Soil microorganisms, 48, 50
Solid culture media, 60
Spirochaeta pallida see Treponema pallidum
Spirochetes, 55
Spiroptera neoplastica, 146
Spores, 56
    tetanus, 19, 20
Stanley, Wendell, 81, 82–4, 131
Streptococcal infection, 40, 57, 72
Streptomyces fradiae, 49
Streptomyces griseus, 48
Streptomyces parvullus, 48
Streptomycin, 2, 47–50, 64
    ear toxicity, 49
    resistance, 48–9
Sugar cane mosaic virus, 83
Sulfa antibiotics see Sulfonamides
Sulfamethoxamole–trimethoprim combination, 53
Sulfanilamide, 40
Sulfonamides, 2, 35, 39, 41, 42
    mechanism of action, 41
    pyrimethamine combinations, 53
Syphilis, 44, 71–5
    clinical stages, 74–5
    fever cabinet treatment, 73
    historical aspects, 73–4
    latent disease, 74
    malariotherapy, 71, 72
    neurosyphilis, 71, 72, 74, 75
    penicillin treatment, 75

T4 helper/inducer cells, 13
    HLA Class II antigen interaction, 33
T8 suppressor/cytotoxic cells, 13
    HLA Class I antigen interaction, 32
    lymphocytic choriomeningitis immune response, 30
    viral infection response, 79
T-cell leukemia/lymphoma virus, 116–17
T-cell receptors, 13
T-cells, 12, 13
    MHC restriction, 30
Tabes dorsalis, 75

Temin, Howard Martin, 115, 116
Tetanospasmin, 20
Tetanus, 15, 19–20, 21, 56
  neonatal disease, 20
  toxin (tetanospasmin), 20, 57
  toxoid immunization, 20
Tetracycline, 2
Theiler, Max, 85, 87
Therapeutic index, 35, 36
Thymidine kinase, 54
Ticks, 134
Tobacco mosaic virus, 81–4
  purification/crystallization, 82–3
  viral characteristics, 81–2
Tomato spotted wilt virus, 83
Toxoid immunization, 16, 20
Toxoplasmosis, 53, 135
Transcription, 77
Translation, 77
Transmissible spongiform encephalopathies,
  124, 125, 126, 128
  prion protein abnormlaities, 129, 131
Transplantation, 31
  MHC/HLA system matching, 33
  rejection, 31, 32, 33
Transport piece, 25
Treponema pallidum, 73, 74, 75
Trichomoniasis, 135
Trimethoprim, 52, 53–4
  sulfamethoxamole combination, 53
Tubercle baccilus see Mycobacterium
  tuberculosis
Tuberculosis, 2, 48, 59–66, 67, 89, 157
  historical aspects, 60
  multi-drug resistance, 60, 64, 65
  spread, 62–3
  symptoms, 63
  treatment, 63–4
  tuberulin skin test, 62, 63
  vaccination, 65–6
Tuberulin test, 62, 63
Tulip breaking virus, 83
Tumor formation, 120
Tumor suppressor genes, 113, 120
Twort, F.W., 103
Typhus, 67–70
  endemic, 70
  epidemic, 68–9
  historical aspects, 68
  reactivation (Brill–Zinsser disease), 70
  transmission by lice, 67, 68–9, 152

Urea synthesis, 84
Urey, H.C., 84
Urinary tract infections, 53
Uterine cervix cancer, 113

Vaccines/vaccination, 11
  acellular, 28
  anthrax, 4
  diphtheria toxoid, 16
  hepatitis A, 99
  hepatitis B, 97–8
  influenza, 81
  malaria, 143
  pertussis (whooping cough), 28
  poliomyelitis, 92–3
  rabies, 1, 2
  smallpox, 1, 4, 88
  tuberculosis, 65–6
  tumor treatment, 79
  viral, production in yeast cells, 108
  yellow fever, 85, 87
Vaccinia virus (cowpox), 1, 2
Varicella-zoster (chickenpox) virus, 78, 90, 94
Variolation, 1
Varmus, Harold, 119–20
Venereal infections, 53
Vinson, 82
Viral encephalitis, 30
Viruses, 2, 29, 35, 77–131
  characteristics, 77–8
  genome, 77
  immune defenses, 79
  in vitro tissue culture, 89, 90, 94
  lysogenic infection, 78–9, 108
  replication, 78, 80
  temporate infection, 79
  tumor production, 79
Vitalism, 84
von Behring, Emil, 15, 16, 19, 20, 21, 28, 145

Wagner-Jauregg, Julius, 71–2, 75
Waksman, Selman, 47–8, 50
Warren, 147
Warts, 113
Welch, Curtis, 17
Weller, Thomas, 89–90, 94
West Nile Virus, 3
White blood cells, 24
Whooping cough (pertussis), 28
Wöhler, Friedrich, 84
Woods, D.D., 41

Yellow fever, 85–8, 134, 152
  clinical features, 85–6
  historical aspects, 86–7
  mosquito control measures, 87–8
  transmission by mosquitoes, 85, 86–7
  vaccine, 85, 87

Zeidler, Othmar, 151
Zinkernagel, Rolf, 29–30, 33